ENVIRONMENTAL REMEDIATION TECHNOLOGIES, REGULATIONS AND SAFETY

FOCUS ON OIL AND GAS

MANAGEMENT, PRODUCTION AND SPILLS

ENVIRONMENTAL REMEDIATION TECHNOLOGIES, REGULATIONS AND SAFETY

Additional books and e-books in this series can be found on Nova's website under the Series tab.

ENVIRONMENTAL REMEDIATION TECHNOLOGIES,
REGULATIONS AND SAFETY

FOCUS ON OIL AND GAS

MANAGEMENT, PRODUCTION AND SPILLS

OLIVIA B. WILDER
EDITOR

snova
New York

Library of Congress Cataloging-in-Publication Data

ISBN: 978-1-53616-325-4

Published by Nova Science Publishers, Inc. † New York

Contents

PREFACE

Chapter 1 examines how BLM's actual costs and potential oil and gas well liabilities have changed for fiscal years 2010 through 2017 and the extent to which BLM has implemented its well and bond review policies. Chapter 2 reviews oil and gas lease suspensions on federal lands managed by BLM and examines the extent of and reasons for such suspensions and the approach BLM uses to monitor the status of lease suspensions.

Chapter 3 describes the distribution of BLM's oil and gas Inspection and Enforcement program's workload and workforce among agency field offices for the most recent 5 years for fiscal years 2012 through 2016 and examines the extent to which BLM conducted internal control reviews in accordance with its July 2012 oversight policy for fiscal years 2013 through 2018.

U.S. oil and natural gas production has increased substantially since 2008. These increases have important policy implications for energy markets, infrastructure, security, and the environment as reported in chapter 4.

Chapter 5 discusses the recent oil market trends and geopolitical factors that have contributed to price escalation since the start of 2018.

Chapter 6 reviews the federal government's response, restoration, and research efforts after the Exxon Valdez and Deepwater Horizon oil spills. This chapter examines how the trustee councils have used the restoration trust funds and the status of restoration and the interagency committee's coordination of oil spill research efforts.

Chapter 1 - In fiscal year 2016, private entities operated about 94,000 oil and gas wells on federal lands overseen by BLM. Once wells cease production, they can become inactive and potentially orphaned if an operator does not perform required reclamation and if an operator's bond is insufficient to cover the expenses. BLM considers oil and gas wells on federal and Indian lands and the associated leased lands as potential liabilities for the federal government because BLM may have to cover the costs of reclaiming well sites. To better manage its potential liabilities, BLM issued well and bond adequacy review policies in 2012 and 2013, respectively. GAO was asked to review how BLM manages its potential oil and gas well liabilities. This chapter examines, among other things: (1) how BLM's actual costs and potential oil and gas well liabilities have changed for fiscal years 2010 through 2017 and (2) the extent to which BLM has implemented its well and bond review policies. GAO analyzed BLM's policies and data and interviewed BLM officials and representatives from stakeholder organizations.

Chapter 2 - Oil and gas leases on federal lands generate billions of dollars in rents and royalty payments for the federal government each year, but these revenues can be reduced if leases are suspended (i.e., placed on hold). Questions have been raised about whether some suspensions, particularly those in effect for more than 10 years, may hinder oil and gas production or adversely affect the use of federal lands for other purposes, such as recreation. GAO was asked to review oil and gas lease suspensions on federal lands managed by BLM. This chapter examines, among other things, (1) the extent of and reasons for such suspensions and (2) the approach BLM uses to monitor the status of lease suspensions. GAO analyzed all data on suspensions in a BLM database and the official lease files for a nongeneralizable sample of 48 leases recorded as suspended in, Montana and Wyoming, which GAO selected based in part on the large number of suspensions these states had. GAO also reviewed BLM documents and interviewed BLM officials.

Chapter 3 - BLM has primary responsibility for managing oil and gas development on federal and Indian lands. To help ensure operator compliance with laws and regulations, BLM administers the Inspection and Enforcement program. Under the program, BLM inspects operators' drilling, production, and plugging activities and can issue various enforcement actions, such as monetary assessments, for violations. GAO was asked to examine key aspects of the Inspection and Enforcement program. This chapter (1) describes the distribution of BLM's oil and gas Inspection and Enforcement program's workload and workforce among agency field offices for the most recent 5 years for which such data were available (fiscal years 2012 through 2016) and (2) examines the extent to which BLM conducted internal control reviews in accordance with its July 2012 oversight policy for fiscal years 2013 through 2018, the most recent period for which such data were available. GAO examined BLM policies, data, and documents; interviewed BLM headquarters, state and field office officials; visited six BLM field offices selected based on their level of resource development activity; and toured oil and gas drilling, production, and plugging sites at three of these six field offices.

Chapter 4 - U.S. oil and natural gas production has increased substantially since 2008. These increases have important policy implications for energy markets, infrastructure, security, and the environment. These complex and interrelated implications have been both positive and negative, depending on perspectives, resulting in a variety of conflicts as well as enormous opportunities. Stakeholders, at times, turn to the courts to address conflicts, such as ensuring that fuel production and consumption comply with federal environmental laws. In the 115th Congress, the House and Senate debated major energy legislation addressing expanding production, pipeline permitting, exports, and energy development on federal land, among other topics.

Chapter 5 - As the 2018 U.S. summer driving season begins, crude oil and gasoline prices have reached their highest levels since 2014. The international crude oil price benchmark (Brent) has increased by approximately 20% since the start of the year, with spot prices reaching just over $80 per barrel in mid-May. As a result, average U.S. retail gasoline prices— including taxes—have reached $2.99 per gallon. Currently, crude oil contributes close to 60% of the price of gasoline. The remaining 40% is from refining, distribution and marketing, and taxes. Prices for other petroleum products such as diesel fuel and kerosene have risen as well. Numerous factors can impact oil and petroleum product prices. Price drivers explored in this In Focus, namely market fundamentals (supply, demand, and stocks) and geopolitical risks, affect the nearly 100 million barrels per day (bpd) global market. However, other factors such as the strength of the U.S. dollar, global spare production capacity, and financial market expectations can also impact price levels.

Chapter 6 - The *Exxon Valdez* and *Deepwater Horizon* oil spills are two of the largest offshore oil spills in U.S. history, causing long-lasting damage to marine and coastal resources. OPA includes provisions to prevent and respond to such oil spills by authorizing (1) federal-state trustee councils that manage billions of dollars from legal settlements and (2) an interagency committee to coordinate oil pollution research, among other things. GAO was asked to review the federal government's response, restoration, and research efforts after the *Exxon Valdez* and *Deepwater Horizon* oil spills. This chapter examines, among other things, (1) how the trustee councils have used the restoration trust funds and the status of restoration and (2) the interagency committee's coordination of oil spill research efforts. GAO reviewed the councils' plans for the funds and how they were used, federal funding of oil spill research by member agencies, and key laws. Also, GAO evaluated the coordination of such efforts against a leading collaboration practice. GAO interviewed members of the trustee councils and the interagency committee.

In: Focus on Oil and Gas
Editor: Olivia B. Wilder

ISBN: 978-1-53616-325-4

Chapter 1

OIL AND GAS WELLS: BUREAU OF LAND MANAGEMENT NEEDS TO IMPROVE ITS DATA AND OVERSIGHT OF ITS POTENTIAL LIABILITIES*

United States Government Accountability Office

ABBREVIATIONS

AFMSS	Automated Fluid Minerals Support System
BLM	Bureau of Land Management
EPAct 2005	Energy Policy Act of 2005
OGOR	Oil and Gas Operations Report
OIG	Office of Inspector General
ONRR	Office of Natural Resources Revenue.

* This is an edited, reformatted and augmented version of United States Government Accountability Office, Report to Congressional Requesters, Publication No. GAO-18-250, dated May 2018.

Why GAO Did This Study

In fiscal year 2016, private entities operated about 94,000 oil and gas wells on federal lands overseen by BLM. Once wells cease production, they can become inactive and potentially orphaned if an operator does not perform required reclamation and if an operator's bond is insufficient to cover the expenses. BLM considers oil and gas wells on federal and Indian lands and the associated leased lands as potential liabilities for the federal government because BLM may have to cover the costs of reclaiming well sites. To better manage its potential liabilities, BLM issued well and bond adequacy review policies in 2012 and 2013, respectively.

GAO was asked to review how BLM manages its potential oil and gas well liabilities. This chapter examines, among other things: (1) how BLM's actual costs and potential oil and gas well liabilities have changed for fiscal years 2010 through 2017 and (2) the extent to which BLM has implemented its well and bond review policies. GAO analyzed BLM's policies and data and interviewed BLM officials and representatives from stakeholder organizations.

What GAO Recommends

GAO is making seven recommendations, including that BLM systematically track the agency's actual reclamation costs and potential liabilities and strengthen its approach to monitoring field offices' implementation of the well review and bond adequacy review policies. BLM agreed with GAO's recommendations.

What GAO Found

GAO's analysis indicates that the Bureau of Land Management's (BLM) actual costs incurred and potential liabilities for reclaiming oil and gas wells have likely increased for fiscal years 2010 through 2017. However, the full extent of the increase is not known because BLM does not systematically track needed data. Based on GAO's analysis of data obtained from 13 of BLM's 33 field offices that manage oil and gas programs, the average annual reclamation cost was $267,600, an increase compared to the $171,500 annual average across all BLM offices that GAO reported in 2010. Similarly, GAO's analysis of BLM data found that the number of known orphaned wells, those that generally have no responsible or liable parties, for all field offices has increased from 144 in 2010 to 219 as of 2017. However, BLM's database that contains information on oil and gas wells on federal and Indian lands does not collect information on costs incurred or on potential liabilities that might result from an increase in the number of orphaned wells. Under federal internal control standards, management should use quality information to achieve the entity's objectives. Without systematically tracking such information, BLM does not have assurance that it has sufficient bonds or financial assurances to cover the costs of reclaiming orphaned wells.

GAO was unable to fully assess the extent to which BLM field and state offices have implemented the agency's policies on reviewing wells and bond adequacy in part because of deficiencies in BLM's monitoring approach. For example, reports BLM headquarters used to monitor field offices' implementation of the policies have limitations. GAO identified discrepancies between the well and bond adequacy review reports that BLM state offices submitted to headquarters and the national summary consolidating states' information. Out of 10 state offices, 3 reported a different number of reviews completed in fiscal year 2016 than what BLM reported in its fiscal year 2016 national summary. Leading practices for monitoring the implementation of agency policies call for taking steps such as collecting and analyzing data on performance indicators. Without strengthening BLM's approach to monitoring, its ability to assess field offices' reviews of all inactive wells and determine the adequacy of all bonds is limited.

May 16, 2018

The Honorable Raúl M. Grijalva
Ranking Member
Committee on Natural Resources
House of Representatives

The Honorable Alan S. Lowenthal
Ranking Member
Subcommittee on Energy and Mineral Resources
Committee on Natural Resources
House of Representatives

The Honorable Jared Polis
House of Representatives

Oil and natural gas resources on federal and Indian lands are an important source of our nation's energy supply. According to the Department of the Interior's Bureau of Land Management (BLM), in fiscal year 2015, the most recent year for which data are available, wells on federal and Indian lands were responsible for providing 11 percent of the natural gas and 7 percent of the oil used in the United States. In fiscal year 2016, private entities operated about 94,000 oil and gas wells on federal lands overseen by BLM.[1] Once these wells cease production, they can become inactive. Inactive wells, which are non-producing wells, have the potential to create physical and environmental hazards if operators fail to reclaim the well sites, which may involve plugging the well, removing structures, and reshaping and revegetating the land around the wells (i.e., returning well sites as close to their original natural conditions as reasonably practical).[2] For example, inactive wells that are not plugged or

[1] Data available from BLM on the total number of wells were limited to federal lands and excluded data on Indian lands.

[2] For the purposes of this chapter, "operator" refers to lessees, owners of operating rights, and operators of an oil or gas operation unless indicated otherwise. We use the term "reclamation" to refer to all of the actions and costs to reclaim a well site, including well

not properly plugged can leak methane or contaminate surface water and groundwater. BLM considers both oil and gas wells on federal and Indian lands as well as the associated leased land as potential liabilities because BLM may have to pay for reclamation if the well operator fails to do so.[3]

BLM regulations and federal laws contain requirements aimed at managing BLM's potential oil and gas well liabilities. For example, to help ensure that operators reclaim well sites, BLM requires operators to provide a bond before drilling operations begin. Operators are required to reclaim well sites before their bonds are released. These bonds may be surety bonds, which are third-party guarantees that operators purchase from private insurance companies, or personal bonds accompanied by a financial instrument, such as a cashier's check. Inactive wells become orphaned if an operator does not perform required reclamation and if the bond is insufficient to cover reclamation expenses and there are no other responsible or liable parties to do so. In these cases, BLM is responsible for completing the reclamation of the well site and uses appropriated funds to perform reclamation.[4] The Energy Policy Act of 2005 (EPAct 2005) directs the Secretary of the Interior to establish a program to reclaim orphaned, abandoned, or idled oil and gas wells on federal lands.[5] The act

---

plugging and surface reclamation, and to restore any lands or surface waters adversely affected by oil and gas operations. BLM differentiates between plugged wells and plugged wells in which the surface has been reclaimed in its data system.

[3] In this chapter, we refer to actual costs BLM incurred to reclaim orphaned wells. We refer to BLM's "potential liabilities" as the costs the agency may incur to reclaim wells that operators fail to reclaim. We do not use "potential liabilities" in the legal or financial accounting sense. We also do not address how BLM recognizes and accounts for environmental liabilities in its financial statements.

[4] BLM regulations do not assign lessee liability for reclamation or decommissioning as "joint and several" as do regulations governing offshore oil and gas operations—that is, each lessee of offshore oil and gas is liable for all decommissioning obligations that accrue on the lease during its ownership, including those that accrued prior to its ownership but had not been performed. The focus of this chapter is on onshore oil and gas wells. For information on the Department of the Interior's management of offshore oil and gas wells, see GAO, *Offshore Oil and Gas Resources: Actions Needed to Better Protect Against Billions of Dollars in Federal Exposure to Decommissioning Liabilities*, GAO-16-40 (Washington, D.C.: Dec. 18, 2015).

[5] Specifically, the act directs the Secretary of the Interior to establish a program to remediate, reclaim, and close such wells. Pub. L. No. 109-58, § 349 *codified as amended at* 42 U.S.C. § 15907. In this chapter we use the term reclamation to include remediation and closing. BLM defines abandoned wells as wells that have been properly plugged and have had final reclamation completed, but have not been approved by the surface managing agency or

requires that the program identify persons providing a bond or other financial assurance and establish a means of recovering the costs of reclaiming wells.

We have previously reported on the actual costs BLM has incurred to reclaim orphaned wells and reported on and made recommendations related to BLM's potential oil and gas well liabilities and the agency's efforts to manage these liabilities. For example, in January 2010, we documented the amount that BLM paid to reclaim orphaned wells from 1988 through 2009 ($3.8 million to reclaim 295 orphaned wells), and identified the number of orphaned wells that BLM had not yet reclaimed (144 orphaned wells).[6] In addition, in February 2011, we found incomplete bond information in BLM's data system, as well as incomplete data on the number of well and bond adequacy reviews.[7] We recommended that BLM develop a comprehensive strategy to better manage potential oil and gas well liabilities. BLM agreed and, partly in response to our recommendations, issued a well review policy in 2012 and a bond adequacy review policy in 2013.[8]

You asked us to review how BLM manages its potential oil and gas well liabilities. This chapter examines (1) how BLM's actual costs incurred to reclaim orphaned wells and potential oil and gas well liabilities have changed, if at all, for fiscal years 2010 through 2017 (since we last reported on these issues); (2) the extent to which BLM has implemented its 2012 well review and 2013 bond adequacy review policies; and (3) BLM officials' and stakeholders' views on what challenges, if any, BLM faces in managing its potential oil and gas well liabilities.

---

wells that have been properly plugged but have not completed final reclamation. An idle well is a well that has been inactive for at least 7 years and has no anticipated beneficial use during the well's lease term.

[6] GAO, *Oil and Gas Bonds: Bonding Requirements and BLM Expenditures to Reclaim Orphaned Wells,* GAO-10-245 (Washington, D.C.: Jan. 27, 2010).

[7] GAO, *Oil and Gas Bonds: BLM Needs a Comprehensive Strategy to Better Manage Potential Oil and Gas Well Liability*, GAO-11-292 (Washington, D.C.: Feb. 25, 2011).

[8] On September 5, 2012, BLM issued Instruction Memorandum 2012-181, which we refer to in this chapter as BLM's well review policy. On July 3, 2013, BLM issued Instruction Memorandum 2013-151, which we refer to in this chapter as BLM's bond adequacy review policy.

To examine how BLM's actual reclamation costs incurred and potential oil and gas well liabilities have changed, we analyzed data from BLM's Automated Fluid Minerals Support System (AFMSS) on oil and gas wells on federal and Indian lands, including inactive and orphaned wells— which represent potential liabilities.[9] AFMSS provides a snapshot of the time that the data are queried, and so AFMSS does not include historical data over time. As such, to examine the number of inactive wells on federal and Indian lands and how long these have been inactive, we combined AFMSS data with data from the Department of the Interior's Office of Natural Resources Revenue (ONRR) Oil and Gas Operations Report's (OGOR) data system. The Department of the Interior requires monthly OGORs from operators. These OGORs document and record the volume of oil and gas produced from wells on federal and Indian lands.[10] To assess the reliability of AFMSS and OGOR data, we reviewed agency documents, met with relevant agency officials, and performed electronic testing. We found the data on the number of inactive wells and how long they have been inactive and the data for the number of wells BLM has identified as orphaned to be sufficiently reliable for our purposes.

However, because AFMSS does not contain information on actual costs incurred to reclaim orphaned wells, we obtained documentation of the actual reclamation costs that 13 selected BLM field offices incurred for fiscal years 2010 through July 2017. To analyze these costs, we reviewed purchase orders and invoices. We also obtained documentation, including spreadsheets with estimated potential reclamation costs that these 13 offices faced as of July 2017.[11] We compared estimated costs provided by the 13 field offices to historical actual costs and determined that overall estimated reclamation costs were sufficiently reasonable for providing a sense of the general magnitude of potential costs, though we did not assess

---

[9] AFMSS is the central database that BLM uses to maintain information on oil and gas activities on federal and Indian lands. Not all data we analyzed covered the entirety of fiscal years 2010 through 2017, due to data limitations and other issues that we describe later in this chapter.

[10] ONRR is responsible for managing and collecting revenues from companies that produce or extract resources on federal leases.

[11] In some cases, officials from the selected field offices continued to provide us with updates on reclamation costs over the course of our review.

the underlying inputs or assumptions used. We selected the 13 BLM field offices because they are responsible for about 80 percent of all oil and gas wells managed by BLM.[12] We also interviewed officials from the 13 selected BLM field offices and the 6 BLM state offices associated with these field offices.[13] Findings from the selected offices cannot be generalized to BLM offices we did not include in our review. We also reviewed documentation provided by BLM and compared BLM's policies and procedures on recording information on actual costs incurred to reclaim orphaned wells and potential liabilities against the information and communication standard outlined in *Standards for Internal Control in the Federal Government.*[14]

To examine the extent to which BLM has implemented its 2012 and 2013 policies for conducting well and bond adequacy reviews, we analyzed BLM's well review and bond adequacy review reports for fiscal year 2016. We also reviewed AFMSS data on wells and bonds as of October 2017. We were unable to fully assess BLM's performance against the directives in the agency's 2012 well review and 2013 bond adequacy review policies due to limited agency data and documentation. We identified data accuracy and consistency concerns with some of the data elements in the agency's well review and bond adequacy review reports as well as some AFMSS data on wells and bonds, which we discuss later in this chapter. In addition, we reviewed documentation from a random, non-generalizable sample of 62 well reviews and 58 bond adequacy reviews, as reported by the 13 selected BLM field offices. Information from our documentation reviews is not generalizable to all BLM field offices but provides illustrative examples of the information contained in BLM's well and bond adequacy reviews. We compared BLM's procedures detailing

---

[12] The 13 selected BLM field offices are as follows: Bakersfield, Buffalo, Casper, Carlsbad, Colorado River Valley, Farmington, Division of Oil and Gas—Great Falls, North Dakota, Oklahoma, Pinedale, Rawlins, Vernal, and Worland. There are 127 BLM field offices, of which 33 manage oil and gas programs. The Division of Oil and Gas—Great Falls changed from a field office to a division in October 2016. For purposes of reporting, we refer to all 13 selected offices as field offices.

[13] The six selected BLM state offices are as follows: California, Colorado, Montana-Dakotas, New Mexico, Wyoming, and Utah.

[14] GAO, *Standards for Internal Control in the Federal Government*, GAO-14-704G (Washington, D.C.: Sept. 10, 2014).

how offices are to count or report a well review and procedures for maintaining data quality against the control activities standard outlined in *Standards for Internal Control in the Federal Government.*[15] We also compared BLM's procedures for monitoring implementation of policy directives against leading practices for monitoring agency policies.[16]

To examine any challenges that BLM faces in managing its potential oil and gas well liabilities, we interviewed officials from 20 BLM offices and interviewed or obtained written responses from representatives from 8 stakeholder organizations knowledgeable about BLM's oil and gas well management, including academic, environmental, industry, and state organizations.[17] In addition, we spoke with knowledgeable officials from ONRR and the Department of the Interior's Office of Indian Energy and Economic Development, Division of Energy and Mineral Development. To identify knowledgeable stakeholders, we conducted a literature search, reviewed previous GAO reports, and obtained recommendations from BLM officials. We then used a technique in which the initial group of BLM officials and stakeholders we interviewed identified additional contacts to interview. From this list, we selected stakeholders who could provide a range of viewpoints. The views of the BLM officials, stakeholders, and other agency personnel we interviewed are not generalizable to BLM officials, similar stakeholders, and other agency personnel we did not interview. Lastly, we compared how BLM identified and managed certain inactive wells, as well as how BLM managed nationwide and statewide bonds, against the control activities standard outlined in *Standards for Internal Control in the Federal Government*;[18] and BLM's resource management practices against certain requirements in EPAct 2005 and leading practices by the Project Management Institute's

---

[15] GAO-14-704G.

[16] We reviewed leading practices for monitoring government programs and projects based primarily on *Standards for Internal Control in the Federal Government*; the GPRA Modernization Act of 2010, Pub. L. No. 111-352, 124 Stat. 3866 (2011); and U.S. Office of Management and Budget, *Uniform Administrative Requirements, Cost Principles, and Audit Requirements for Federal Awards*, 2 C.F.R. part 200.

[17] We interviewed officials from the 13 selected BLM field offices, the 6 state offices associated with these 13 field offices, and BLM headquarters, for a total of 20 BLM offices.

[18] GAO-14-704G.

The Standard for Program Management.[19] (App. I lists the BLM offices and knowledgeable stakeholders we interviewed and provides additional information on our scope and methodology).

We conducted this performance audit from November 2016 to May 2018, in accordance with generally accepted government auditing standards. Those standards require that we plan and perform the audit to obtain sufficient, appropriate evidence to provide a reasonable basis for our findings and conclusions based on our audit objectives. We believe that the evidence obtained provides a reasonable basis for our findings and conclusions based on our audit objectives.

BACKGROUND

This section provides information on (1) BLM headquarters, state, and field offices; (2) the lifecycle of oil and gas wells; (3) BLM's bonding regulations; and (4) BLM's 2012 well review and 2013 bond adequacy review policies.

BLM Headquarters, State, and Field Offices

BLM is responsible for issuing leases for private entities to develop oil and gas resources on and under roughly 700-million acres of (1) BLM land, (2) other federal agencies' land, and (3) private land where the federal government owns the mineral rights. According to BLM, approximately 32-million acres were leased for oil and gas operations at the end of fiscal year 2015. BLM also oversees oil and gas operations on 56-million acres of Indian lands. BLM administers its programs through its headquarters office in Washington, D.C.; 12 state offices; 38 district offices; and 127 field offices.

---

[19] Project Management Institute, Inc., *The Standard for Program Management*, Fourth Edition (Newtown Square, PA: 2017). PMI® is a registered mark of the Project Management Institute, Inc.

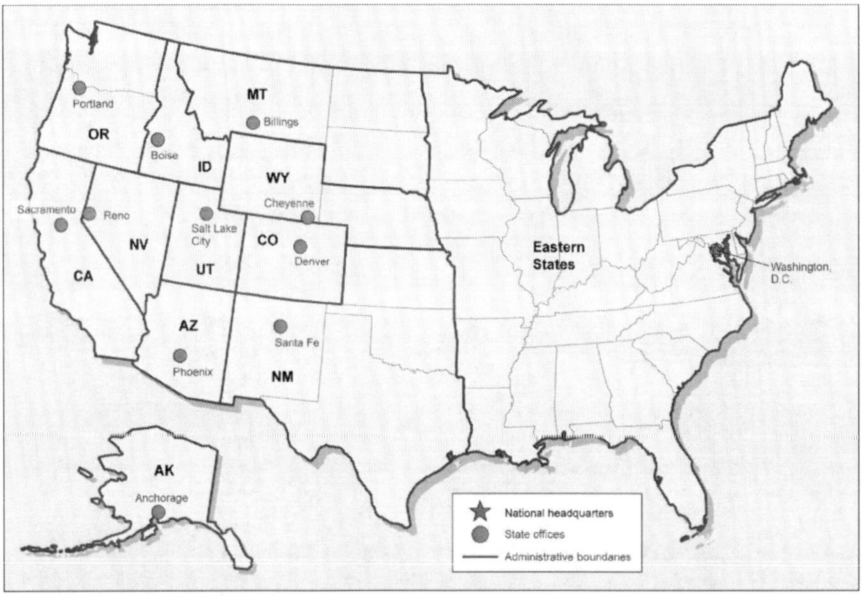

Sources: GAO analysis of Bureau of Land Management data; Map Resources (map). GAO-18-250.

Figure 1. Boundaries of the 12 Bureau of Land Management State Offices.

Of these, 10 state offices and 33 field offices manage oil and gas programs, and these are located primarily in the Mountain West, the center of much of BLM's oil and gas development. BLM headquarters develops guidance and regulations for the agency, and the state, district, and field offices manage and implement the agency's programs. Because BLM has few acres of land in the eastern half of the United States, the Eastern States State Office, in Washington, D.C., is responsible for managing land in 31 states, and the remaining state offices generally conform to the boundaries of one or more states. Figure 1 shows the boundaries of the 12 BLM state offices.

Lifecycle of Oil and Gas Wells

Once operators obtain federal oil and gas leases and drill wells, those wells can be actively producing, inactive, or reclaimed. An orphaned well is a well that BLM determined has no responsible or liable party and for which there is insufficient bond coverage for reclamation.[20] This situation may occur, for example, when an operator has declared bankruptcy. Shut-in and temporarily abandoned wells are examples of types of inactive wells that can become orphaned.

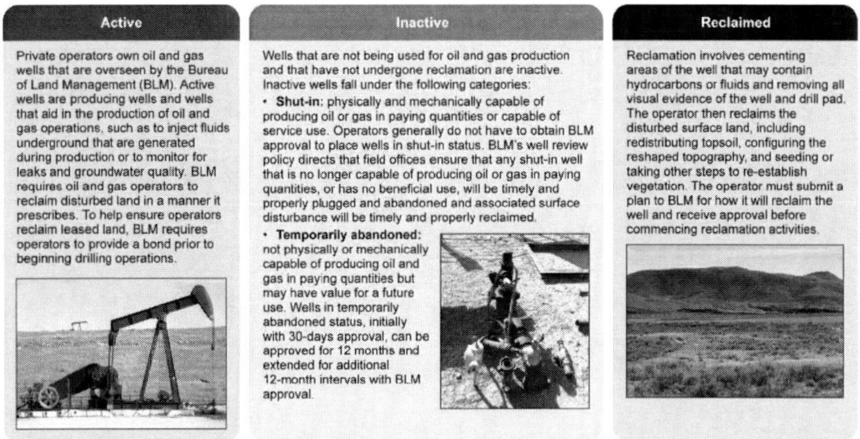

Source: GAO analysis of BLM information; Department of Energy (left photo); GAO (center and right photos). GAO-18-250.

Figure 2. Lifecycle of Oil and Gas Wells Overseen by the Bureau of Land Management (BLM).

Shut-in wells are physically and mechanically capable of producing oil or gas in paying quantities or capable of service use. For example, an operator may put a well in shut-in status if it has not been connected to a sales line or the line is too far away and it is not economical to connect to

[20] According to BLM, field offices typically go through several steps to determine that a well has no responsible or liable party and to designate a well as orphaned, including issuing *Incidents of Noncompliance* to the operator, notifying the operator of proposed civil penalties, and terminating the lease. If the operator fails to respond, the field office is to issue the same notifications to the current record title owners and require them to either comply or reclaim the well sites. The field office is to also determine whether the lease is capable of producing hydrocarbons in paying quantities. The 2012 well review policy defines production in paying quantities as covering the costs to operate the well and market the products. The phrase "market the products" is further defined as the well's processing, measurement, and transportation costs that a prudent operator would be expected to incur in order to sell the products. The process to designate a well as orphaned can take up to several years.

at this time. Temporarily abandoned wells are another type of inactive well that is not physically or mechanically capable of producing oil or gas in paying quantities but that may have value for a future use. Figure 2 depicts the lifecycle of oil and gas wells overseen by BLM.

BLM's Bonding Regulations

The Mineral Leasing Act of 1920, as amended, requires that federal regulations ensure that an adequate bond is established before operators begin preparing land for drilling to ensure complete and timely reclamation of the land. Accordingly, BLM regulations require operators to submit a bond to ensure compliance with all of the terms and conditions of the lease, including, but not limited to paying royalties, plugging wells, and reclaiming disturbed land.[21] BLM regulations generally require operators to have one of the following types of bond coverage:[22]

- individual lease bonds, which cover all of an operator's wells under one lease, and the minimum amount is set at $10,000;[23]
- statewide bonds, which cover all of an operator's leases in one state, and the minimum amount is set at $25,000;[24] or
- nationwide bonds, which cover all of an operator's leases in the United States, and the minimum amount is set at $150,000.[25]

---

[21] 43 C.F.R. § 3104.1(a).

[22] Other bonds include both unit operator bonds that cover all operations conducted on leases within a specific unit agreement, and bonds for leases in the National Petroleum Reserve in Alaska. Unit agreements refer to multiple lessees who unite to adopt and operate under a single plan for the development of any oil or gas pool, field, or like area. The amount of a unit operator bond is determined on a case-by-case basis by BLM officials, and the minimum amount of a National Petroleum Reserve in Alaska bond is set in regulation—not less than $100,000 for a single lease or not less than $300,000 for a reserve-wide bond (submitted separately or as a rider to an already existing nationwide bond).

[23] An individual lease bond posted by a lessee may cover all operators on a lease. Otherwise, each operator on a lease must provide a separate bond covering just the wells operated by that operator. As we previously reported, according to BLM officials, most leases have only one operator. See GAO-11-292.

[24] A statewide bond posted by a lessee can cover all well operators with the consent of the company providing the bond.

BLM can accept two types of bonds: surety bonds and personal bonds. A surety bond is a third-party guarantee that an operator purchases from a private insurance company approved by the Department of the Treasury. The operator is required to pay a premium to the surety company to maintain the bond. These premiums can vary depending on various factors, including the amount of the bond and the assets and financial resources of the operator. If operators fail to reclaim the land they disturb, the surety company can either pay BLM the amount of the bond to help offset reclamation costs, or in some circumstances, BLM may allow the surety company to perform the required reclamation. A personal bond must be accompanied by one of the following financial instruments:

- certificates of deposit issued by a financial institution whose deposits are federally insured, granting the Secretary of the Interior authority to redeem it in case of default in the performance of the terms and conditions of the lease;
- cashier's checks;
- certified checks;
- negotiable Treasury securities, including U.S. Treasury notes or bonds, with conveyance to the Secretary of the Interior to sell the security in case of default in the performance of the lease's terms and conditions; or
- irrevocable letters of credit that are issued for a specific term by a financial institution whose deposits are federally insured and meet certain conditions.

If operators fail to reclaim the land they disturb, BLM will redeem the certificate of deposit, cash the check, sell the security, or make a demand on the letter of credit to pay the reclamation costs.

---

[25] A nationwide bond posted by a lessee can cover all well operators with the consent of the company providing the bond.

BLM's 2012 Well Review and 2013 Bond Adequacy Review Policies

In response to our previous recommendations that BLM develop a comprehensive strategy to improve monitoring agency performance in conducting well reviews and bond adequacy reviews, BLM issued a 2012 well review policy and a 2013 bond adequacy review policy. These policies contain directives for conducting reviews when wells and bonds meet certain criteria. The well review policy directs:

- that field office officials evaluate every shut-in well at least once every 5 years;
- that field office officials review all wells that have been inactive for 25 years or longer and that have no anticipated beneficial use by March 29, 2013;
- that if field office officials determine that there are wells that are not capable of producing oil or gas in paying quantities or have no beneficial use, officials are to send the operator a written order directing the operator to demonstrate that these wells are capable of producing oil or gas in paying quantities or have a future beneficial use, or the operator is to submit plans to reclaim the wells;
- that each state office submit to BLM headquarters a consolidated annual report recording well reviews; and
- that the annual report identify the leases that were reviewed and the wells that were reviewed on each lease, and describe what follow-up action the field office official conducting the review performed.

The bond adequacy review policy directs:

- that field offices perform bond adequacy reviews on all bonds at least once every 5 years or whenever a bond review is warranted;

- that field offices verify and tie all federal wells to their appropriate bond number and enter bond information and bond adequacy review data into AFMSS;
- that field offices perform adequacy reviews on all bonds using specific instructions and a worksheet that assigns points for three risk factors: (1) status of wells covered by the bond (share of inactive wells, deep wells, and wells with marginal production); (2) operator-specific compliance history; and (3) reclamation stewardship diligence;
- that if the field office official performing the review determines that the bond amount is insufficient, the official is to take the necessary steps to determine the appropriate bond amount and increase the bond;
- that if the bond being reviewed is a statewide or nationwide bond, field offices are to review the wells within their field office jurisdiction; and
- that each BLM state office with an oil and gas program submit a semiannual bond adequacy review report to BLM headquarters.

BLM'S ACTUAL RECLAMATION COSTS AND POTENTIAL OIL AND GAS WELL LIABILITIES HAVE LIKELY INCREASED, BUT THE AGENCY DOES NOT SYSTEMATICALLY TRACK THESE

BLM's actual costs incurred to reclaim orphaned wells and potential liabilities have likely increased for fiscal years 2010 through 2017 based on our analysis of available information. Precisely how the agency's actual reclamation costs and potential liabilities have changed is unclear because BLM does not systematically track them at an agency-wide level. BLM headquarters officials we interviewed told us that they did not have any information on actual costs incurred to reclaim orphaned wells and stated that BLM's data systems were not designed to track incurred reclamation costs. In addition, AFMSS provides a snapshot of orphaned wells as identified at the time that the data are queried and does not provide data for prior time periods.

Because BLM headquarters does not record actual reclamation costs incurred at an agency-wide level, we requested documentation for the reclamation costs incurred by 13 selected BLM field offices for fiscal years 2010 through July 2017.[26] This documentation identified about $2.1 million in reclamation costs incurred over this period, or an average of about $267,600 per year by these 13 field offices. We estimate that total actual reclamation costs for all field offices are likely to be higher than this amount as other field offices may have also reclaimed orphaned wells during this period. In January 2010, we found that, for all field offices across the agency, BLM spent about $3.8 million from fiscal years 1988 through 2009, or an average of about $171,500 per year.[27] Comparing the average costs incurred by the 13 selected field offices to the data we previously reported demonstrates that actual total reclamation costs incurred have likely increased since 2010.

In addition to actual costs increasing, potential liabilities are also likely to have increased, though BLM does not systematically track information on potential liabilities that might result from an increase in the number of orphaned wells. Potential liabilities include costs that the agency may incur to reclaim wells that operators fail to reclaim. We believe these costs have also increased because the number of known orphaned wells on federal and Indian lands managed by BLM has increased. We identified changes in the number of known orphaned wells since we last reported on this matter in January 2010. In January 2010, we found that BLM had identified and was managing 144 orphaned wells.[28] Over half of those 144 wells (75) were still identified in AFMSS as orphaned as of July 2017, and the total

---

[26] Officials from selected field offices provided documentation on actual reclamation costs, including purchase orders and invoices. In some cases, officials from the selected field offices continued to provide us with updates on reclamation costs over the course of our review.

[27] GAO-10-245.

[28] GAO-10-245. In that report, we found that BLM's reclamation cost estimates were not available for all of the wells it had yet to reclaim, but BLM field office officials completed reclamation cost estimates of approximately $1.7 million for 102 of the 144 orphaned wells.

number of identified orphaned wells on federal lands had increased from 144 to 219.[29] Also, BLM officials from the 13 selected field offices identified about $46.2 million in estimated potential reclamation costs associated with orphaned wells and inactive wells that officials deemed to be at risk of becoming orphaned.[30]

Also concerning potential liabilities, our analysis of AFMSS data and OGOR production data through September 2016 found that BLM managed about 15,600 inactive wells, of which over 1,000 were inactive for 25 years or more. In contrast, a document provided to us by BLM headquarters indicates 325 wells had been inactive for 25 years or more as of around 2017. This document summarizes data from AFMSS queries conducted by BLM field and state offices at various times from 2013 through 2014 and queries conducted at various times from 2016 through 2017. BLM officials told us that this difference could be because AFMSS reports sometimes return conflicting data since the reports draw from current and historical statuses of wells from both AFMSS and OGOR. We combined AFMSS and OGOR data to identify the number of inactive wells because although BLM records the total number of wells on federal lands over time—a rough indicator of how potential reclamation costs may change—the agency does not systematically record more specific types of wells that may be at higher risk of becoming orphaned, such as inactive wells or wells that have been inactive for 25 years or more.

---

[29] As of July 2017, there were 23 wells identified as orphaned on Indian lands, in addition to the 219 orphaned wells on federal lands, for a total of 242 orphaned wells. In January 2010, the number of orphaned wells we reported on did not include wells on Indian lands. According to an official we interviewed for this chapter from the Department of the Interior's Division of Energy and Mineral Development, orphaned wells on Indian lands could, in certain instances, represent a potential liability to the federal government.

[30] Officials from one field office told us that they estimated potential reclamation costs based on various factors, such as depth of the well, location, and seeding requirements. To assess the reasonableness of estimated reclamation costs, we reviewed estimates provided by officials from the selected field offices and compared those to historical actual costs that we previously reported in January 2010. We determined the overall estimated reclamation costs were sufficiently reasonable for providing a sense of the general magnitude of potential costs, though we did not assess the underlying inputs or assumptions used. Total potential reclamation costs for all field offices are likely to be higher than the costs for these 13 field offices.

Moreover, we identified inconsistencies between the data and the document provided to us by BLM headquarters summarizing the data. For example, BLM's summary document did not include one state office, even though the data include that state office as having two wells that were inactive for 25 years or more in 2014. BLM's summary document states that there had been a reduction in the number of wells that were inactive for 25 years or more between the times of the two data queries. However, because BLM does not systematically track the number of inactive wells, in particular those wells that are at high risk of becoming orphaned, the agency does not know how its potential liabilities may be changing. These liabilities include wells inactive for 25 years or more.

Although we were unable to determine the full extent of the increase in BLM's potential liabilities because BLM does not have the data needed for such an analysis, other factors also suggest such an increase. For example, there has been an increase in oil and gas development on federal lands, and therefore, there is the potential for an increase in the total number of wells on federal lands at risk of becoming orphaned and needing to be reclaimed in the future. BLM's portfolio of oil and gas wells on federal lands has changed over the years, based on overall trends in the oil and gas industry.[31] According to AFMSS data provided by BLM, the total number of wells on federal lands that are capable of production increased along with rising oil and gas prices, from about 89,600 wells in fiscal year 2010 to peaking to about 94,800 wells in fiscal year 2014. As oil and gas prices declined starting in 2014, the total number of wells capable of production also declined to about 94,100 wells in fiscal year 2016.

In addition, declining oil and gas prices (by nearly half from 2010 through 2017) have placed financial stress on oil and gas operators, thereby increasing bankruptcies and the risk of wells becoming orphaned. For example, coalbed methane—natural gas extracted from coal beds—was economical to produce when natural gas prices were higher and thousands

---

[31] AFMSS data provided by BLM on oil and gas development and the total number of wells were limited to federal lands and excluded data on wells on Indian lands.

of coalbed methane wells were drilled on federal lands.[32] However, coalbed methane production has declined because the spread of shale gas production has driven down natural gas prices. Officials we interviewed in one BLM field office told us that the drop in natural gas prices contributed to an increasing number of bankruptcies for operators of coalbed methane wells. Our analysis of AFMSS data suggests that there were thousands of inactive coalbed methane wells as of October 2017. To the extent that market conditions remain unfavorable for coalbed methane production, BLM's potential future reclamation costs may increase if any operators of these wells go bankrupt or are otherwise unwilling or unable to pay the full costs of reclamation, leaving these wells orphaned.

According to federal internal control standards, management should use quality information, which should be complete, to achieve the entity's objectives.[33] However, BLM does not systematically or comprehensively track the agency's actual costs incurred to reclaim orphaned wells and the information necessary to determine potential liabilities, including indicators of potential future reclamation costs, such as the number of inactive wells, orphaned wells, and estimates of reclamation costs for orphaned wells. BLM headquarters officials said that they sometimes check AFMSS to see how many orphaned wells there are, but without doing so systematically and recording the results of these checks, it is not possible to determine how the agency has been making progress in managing the number of orphaned wells. EP Act 2005 requires that the costs of reclaiming orphaned wells be recovered from persons or entities providing a bond or other financial assurance. Without systematically and comprehensively tracking actual reclamation costs incurred and the information necessary to determine potential liabilities including the numbers of orphaned wells and inactive wells over time, BLM cannot ensure that it has sufficient bond coverage or other financial assurances to

---

[32] In coalbed methane formations, gas is extracted through a process to reduce pressure called dewatering. As water is pumped out of the coal seams, reservoir pressure decreases, allowing the natural gas to release (desorb) from the surface of the coal and flow through natural fracture networks into the well. The separation of the gas from the water as well as the disposal of the water may be costly, making this process relatively expensive compared to natural gas production from formations containing less water.

[33] GAO-14-704G.

minimize the need for taxpayers to pay for the costs of reclaiming orphaned wells.

THE EXTENT TO WHICH BLM IMPLEMENTED ITS WELL REVIEW POLICY AND BOND ADEQUACY REVIEW POLICY DIRECTIVES IS UNCLEAR

The extent to which BLM has implemented its well review policy and bond adequacy review policy is unclear. Specifically, we were unable to fully assess the extent to which BLM's field and state offices have implemented directives included in these policies because of inconsistent well review information, inaccurate well and bond data in AFMSS, and inadequate monitoring of well and bond policies' implementation.

Inconsistent Well Review Information

We were unable to fully assess the extent to which BLM implemented some directives in the well review policy because the well review information reported by field offices differed across the agency. For example, officials we interviewed at the 13 selected BLM field offices had different understandings of what specific actions constitute a well review, and therefore differed in their understanding of which wells were to be included in the annual reports for documenting well reviews. Specifically, officials from 11 out of 13 selected field offices told us that a well review consisted of actions—such as reviewing a well's status, conducting a physical inspection, and providing additional notices or letters to the well operator when a well is inactive. Officials in 2 other field offices told us that while they conduct similar actions, they consider the sole action of correcting data on a well's status to constitute a well review.[34] For

[34] In addition, we found that the way state offices reported well reviews varied between reporting at the well completion level and reporting at the larger well level, resulting in inconsistent

example, a BLM official told us that one reported well review was conducted on a well that had been reclaimed in 1986 but that was not noted in AFMSS. The official told us that following this well review, they corrected the well status in AFMSS and noted that this well should not have been on the list of wells to review. While correcting well data helps improve the accuracy of AFMSS, when some offices count such corrections as well reviews and others do not, this variance results in inconsistent information in BLM's annual well review reports.

Such inconsistencies in what counts as a well review may be the result of a lack of clarity in BLM's well review policy that does not specify what constitutes a well review. Unlike the bond adequacy review policy, which provides instructions to field offices on how to conduct a bond adequacy review and directs field offices to use a specific worksheet to calculate bond adequacy, the well review policy does not contain specific instructions on what actions field offices are to take to conduct a well review, such as how to count reviews or report them. A January 2018 report by the Department of the Interior's Office of Inspector General (OIG) similarly found that BLM's well review policy does not specifically outline how to conduct and document reviews of shut-in wells (shut-in wells, as noted earlier, are inactive wells that are physically and mechanically capable of producing oil or gas in paying quantities or capable of service use).[35] Under federal standards for internal control,

---

reporting. BLM officials explained that a well with multiple completions is equipped to produce oil or gas separately from more than one zone. A well may be shut-in or plugged in one zone but re-entered or worked over to open up production from another zone. The two zones could be totally independent of each other. Thus some wells that are producing from multiple completions are erroneously reported as shut-in if one zone is inactive. BLM officials told us that a well review is conducted at the well level, not at the completion level. BLM officials also told us that any well with a producing zone should be considered as a producing well.

[35] Department of the Interior, Office of Inspector General, *Bureau of Land Management's Idle Well Program*, Report No. 2016-EAU-061 (Washington, D.C.: Jan. 17, 2018). The OIG found that BLM management could not ensure its field offices were conducting reviews appropriately for managing their idle well inventories or taking some of the steps necessary to help reduce those inventories. The OIG recommended that BLM develop and implement guidance or update its policy regarding how to conduct and document reviews of shut-in wells. BLM concurred with the recommendation. In addition, the OIG found that unlike BLM's Inspection and Enforcement strategy, which details how to prioritize wells for inspection, BLM has no clear strategy for prioritizing and conducting well reviews,

management should design control activities to achieve objectives and respond to risks; such activities include appropriate documentation of internal control in management directives, administrative policies, or operating manuals. Without developing and communicating specific instructions outlining what actions constitute a well review for annual-reporting purposes, BLM cannot have reasonable assurance that its field offices are conducting and reporting on well reviews in a consistent manner.

Inaccurate Well and Bond Data in AFMSS

Our ability to assess the extent to which BLM implemented its well review and bond adequacy review policies was impeded by inaccuracies in certain AFMSS data. BLM officials told us that some of the data in AFMSS on wells and bonds were not reliable. For example, BLM officials told us that there may be discrepancies between the bonds listed in AFMSS and the bonds listed in the Bond and Surety System, which is BLM's official database for all oil and gas bonds. Officials told us that bonds may be missing from AFMSS because BLM field offices are responsible for manually entering the bond number from the Bond and Surety System into AFMSS.[36]

In addition, AFMSS data we reviewed contained other inaccuracies. Specifically, the data we reviewed contained future dates for when wells were completed, or capable of production, when some wells last changed statuses, and when some well reviews were reportedly conducted. BLM officials told us that AFMSS allows users to enter future dates, which can result in inaccurate data. Having inaccurate dates for wells' statuses and wells' reviews is problematic because it means it is not possible to assess

resulting in inconsistencies in how field offices prioritize well reviews. The OIG recommended that BLM develop and implement a well review strategy for inactive wells consistent with BLM's Inspection and Enforcement strategy. BLM also concurred with this recommendation.

[36] According to our analysis of AFMSS data provided by BLM officials in December 2017, about 10 percent of wells did not have an associated bond.

whether reviews are being conducted as directed by BLM policy. For example, BLM's well review policy directs field offices to review each shut-in well every 5 years. BLM's performance against this directive cannot be assessed without reliable information on when wells become shut-in and when well reviews are conducted.

In written responses to our request for information, BLM officials stated that AFMSS has some edit checks, but the accuracy of the data entered into AFMSS is dependent on field office officials responsible for data entry. BLM officials stated that AFMSS has some electronic safeguards, such as certain number fields only accepting numbers. In addition, AFMSS has dropdown menus and checkboxes to narrow the parameters of certain data being entered. However, there are no edit checks to prevent field offices from inputting future status dates. In addition, BLM's data administration and management handbook establishes that data stewards are to, among other things, establish target quality levels, data quality plans (including audits and other quality assurance steps), and certify the quality of the data. BLM officials stated that they have national level AFMSS data stewards and information-technology data stewards. However, BLM officials stated that the agency has not defined AFMSS target quality levels and did not provide any data quality plans. Officials stated that BLM headquarters conducts annual data reviews and will periodically review sample well files to detect data inconsistencies and errors. In addition, BLM officials stated that field offices are responsible for certifying the accuracy of the data they enter into AFMSS, and BLM headquarters is responsible for providing oversight. However, BLM headquarters officials did not provide documentation of any data certifications or data reviews, raising concerns over the extent of this oversight.

Under federal standards for internal control, management should design control activities, including control activities used in information processing, to achieve objectives and respond to risks. Examples of such control activities include:

- conducting edit checks of data entered,
- accounting for transactions in numerical sequences, and
- comparing file totals with control accounts.

Without taking steps to improve AFMSS data quality, such as by conducting more edit checks and having data stewards certify the quality of the data, BLM cannot have reasonable assurance that management has the accurate information it needs to track whether field offices are conducting well and bond adequacy reviews as intended.

In its January 2018 report, the OIG found similar issues related to the accuracy of AFMSS data.[37] Specifically, the OIG found that AFMSS data were unreliable due to inaccurate well status information. The OIG also found that BLM officials update AFMSS manually during a well review or as needed, as opposed to automating the data, meaning that information about the status of individual wells in AFMSS and data used for BLM's annual well report are not timely. The OIG recommended that BLM develop and implement a quality control process to identify inaccurate or incomplete data in AFMSS. BLM concurred with this recommendation.

Inadequate Monitoring of Well and Bond Policies' Implementation

BLM headquarters has taken some actions to monitor the implementation of its well and bond adequacy review policies across the agency, but its efforts have been limited, and the agency cannot ensure that its policy directives have been fully implemented. For example, BLM headquarters officials told us that headquarters relies on national well review and bond adequacy review reports to monitor the extent to which field offices are conducting well and bond adequacy reviews. These well and bond adequacy review reports provide some information on how BLM

[37] Department of the Interior, Office of Inspector General, *Bureau of Land Management's Idle Well Program*, Report No. 2016-EAU-061 (Washington, D.C.: Jan. 17, 2018).

field offices conducted their reviews during a given year, but the reports as previously mentioned above have data limitations and do not consistently record a field office's progress in meeting the policies overall. For example, annual well review reports list the wells field offices reviewed in a given year, but do not compare this statistic to a list of the wells that each field office should have reviewed. Similarly, field offices' bond adequacy review reports list the bonds that the field offices reviewed in a given year. However, the reports do not compare the bonds reviewed to a list of bonds each field office should have reviewed.

In addition, our analysis of 58 selected bonds reported as reviewed across the 13 selected field offices found that 4 bonds—about 7 percent—were not reviewed, even though field offices had reported that they had conducted the reviews.[38] The bond adequacy review policy directs field offices to review all bonds once every 5 years or whenever a bond review is warranted. Therefore, the bond adequacy review reports on their own provide insufficient information for BLM headquarters to monitor progress about whether field offices are fully implementing the directive.

We also identified discrepancies between the annual well review and semi-annual bond adequacy review reports that state offices submitted to BLM headquarters and the information in headquarters' national summary, which consolidates the state office information. These discrepancies limit the usefulness of the national summary for monitoring the extent to which field offices are conducting well and bond adequacy reviews as directed by the policies. For example, 3 out of 10 state offices reported a different number of bond adequacy reviews completed in their fiscal year 2016 state reports than what was reported in BLM's fiscal year 2016 national report.[39] Similarly, 6 out of 9 state offices reported a different number of completed

---

[38] For one of the four bonds, field office officials told us that they verified that all required wells and liabilities were associated with the new bond, and that this review was not conducted in accordance with the bond adequacy review policy. We therefore did not consider it reviewed as part of our analysis.

[39] In addition, BLM headquarters officials did not provide semi-annual bond adequacy review reports covering the first and second quarters as directed by the policy for any of the six selected state offices. Six out of the six selected state offices provided end-of-year bond adequacy review reports that, according to headquarters officials, include data for the entire year selected.

well reviews in their fiscal year 2016 state report than what was reported in BLM's fiscal year 2016 national report.[40]

Similarly, the OIG's January 2018 report found that BLM can only report its progress in reviewing wells that have been inactive for 25 years or more by using field office spreadsheets, coupled with AFMSS data.[41] The report stated that using spreadsheets and AFMSS data have made it difficult, however, for BLM to demonstrate proper oversight. BLM's headquarters officials had to ask state office officials how many wells had been reviewed and then had to summarize those results in a spreadsheet. The OIG recommended that BLM monitor and track reviews of shut-in wells in a management system. BLM concurred and stated that AFMSS and an update to AFMSS that is under development were the appropriate databases for monitoring and tracking well reviews.

Overall, we found that BLM's current approach to monitoring the agency's progress in implementing its well and bond adequacy review policies has been limited. We reviewed leading practices for monitoring the implementation of agency policies.[42] These practices call for, among other things: (1) periodically collecting and analyzing data on performance indicators, (2) establishing procedures for ensuring the quality of data on performance indicators, (3) documenting that monitoring plans were executed, and (4) considering performance information in making management decisions. Without taking actions to strengthen its approach to monitoring, such as collecting and analyzing data on performance indicators and ensuring the quality of those data, BLM's ability to assess the extent to which field offices are reviewing all inactive wells and determining the adequacy of all bonds is limited.

---

[40] BLM officials did not provide a state report for Idaho.

[41] Department of the Interior, Office of Inspector General, *Bureau of Land Management's Idle Well Program*, Report No. 2016-EAU-061 (Washington, D.C.: Jan. 17, 2018).

[42] We reviewed leading practices for monitoring government programs and projects based primarily on *Standards for Internal Control in the Federal Government*; the GPRA Modernization Act of 2010, Pub. L. No. 111-352, 124 Stat. 3866 (2011); and U.S. Office of Management and Budget, *Uniform Administrative Requirements, Cost Principles, and Audit Requirements for Federal Awards*, 2 C.F.R. part 200.

Agency Officials and Stakeholders Identified Several Challenges BLM Faces in Managing Its Potential Oil and Gas Well Liabilities

According to BLM officials and stakeholders we interviewed, BLM faces several challenges in managing its potential liabilities.[43] In particular, BLM officials and stakeholders told us that one challenge in managing BLM's potential liabilities was identifying and managing shut-in wells and preventing them from becoming orphaned. Another challenge identified was limited resources and competing priorities in reclaiming orphaned wells. Other challenges to managing BLM's potential liabilities include difficulties in reviewing nationwide bonds, minimum bond amounts, and operators' unresponsiveness.

BLM Faces Challenges Identifying and Managing Shut-in Wells and Preventing Them from Becoming Orphaned

BLM officials from 6 of the 20 BLM offices—including headquarters and selected state and field offices—and 2 of the 10 stakeholders told us that one of the challenges that BLM faces in managing its potential liabilities is identifying and managing shut-in wells. As previously mentioned, shut-in wells are inactive wells that are physically and mechanically capable of producing oil or gas in paying quantities or capable of service use. Since shut-in wells may become orphaned and therefore involve BLM resources to reclaim, identifying and managing

[43] We interviewed officials from 13 selected BLM field offices that manage the largest number of oil and gas wells, representing about 80 percent of all oil and gas wells managed by BLM. We also interviewed BLM officials from the 6 state offices associated with these 13 field offices, and BLM headquarters, for a total of 20 BLM offices. We also interviewed or received written responses from representatives of 8 stakeholders who were knowledgeable about BLM's oil and gas well management, including academic, environmental, industry, and state organizations. In addition, we spoke with officials from 2 agencies within the Department of the Interior. For purposes of this chapter, we refer to these 10 stakeholders and Department of the Interior officials collectively as stakeholders.

them is a way for BLM to manage its potential liabilities. BLM's 2012 well review policy directs field offices to review all shut-in wells on federal and Indian lands every 5 years and to ensure that shut-in wells no longer capable of production are reclaimed. However, operators are generally not required to notify BLM when they place a well in shut-in status. As a result, officials noted that it is difficult for field offices to identify all shut-in wells in order to review them. Officials from one field office told us that identifying when a well becomes shut-in is challenging unless inspectors are able to physically find the well.

Even when wells have been identified to BLM as shut-in, some BLM officials at selected field offices said that they have few policy tools to manage shut-in wells. In reviewing the well review policy, we found that it contains certain directives for wells that are temporarily abandoned, including that an operator is to conduct well integrity testing prior to placing a well in temporarily abandoned status and a 30-day limit for how long operators can place wells in temporarily abandoned status without receiving BLM approval. However, the policy contains no similar directives related to testing or limited time frames for placing wells in shut-in status. As a result, BLM may be unable to identify and reduce its inventory of shut-in wells, including wells that have been in shut-in status for an extended period of time.

In its January 2018 report, the OIG similarly found that the well review policy does not provide field offices the leverage to make an operator conduct integrity testing since the policy does not have instructions on the method, frequency, and way to proceed with a notice or order.[44] Without having these test results available to them, the report found that BLM staff cannot be certain that an inactive well is environmentally sound and capable of production. The report recommended that BLM develop and implement guidance or update the well review policy to require integrity testing on inactive wells at specific periods.

Strengthening the identification and management of shut-in wells could be particularly helpful in managing BLM's potential liabilities

---

[44] Department of the Interior, Office of Inspector General, *Bureau of Land Management's Idle Well Program*, Report No. 2016-EAU-061 (Washington, D.C.: Jan. 17, 2018).

because such wells have represented a large portion of orphaned wells. According to our analysis of AFMSS data, 138 of the 242 orphaned wells BLM manages were in shut-in status prior to becoming orphaned. Moreover, one of these wells had been in shut-in status since 1926. BLM's Colorado and New Mexico state offices have taken steps to address the challenges associated with shut-in wells becoming orphaned. For example, in September 2016, BLM's New Mexico state office issued a policy that directed operators to obtain BLM's approval in order to place a well in shut-in status for more than 90 days and directed the operator to conduct periodic testing to verify that wells that have been inactive for more than 12 consecutive months remain capable of production.[45] Under federal standards for internal control, management should design control activities—such as by clearly documenting internal control in management directives, administrative policies, or operating manuals—to achieve objectives and respond to risks.[46] Without providing greater specificity in current policy or new supplemental guidance to all BLM field offices on how to identify and manage shut-in wells, the agency is at an increased risk of having unidentified shut-in wells, and wells that remain in shut-in status for extended periods of time, leading to increased potential liabilities if such wells become orphaned.

BLM Faces Challenges Related to Limited Resources and Competing Priorities

BLM officials and stakeholders told us that one of the challenges BLM faces in managing its potential liabilities is limited resources, including staff and funding, and competing priorities. Specifically, officials from 14

---

[45] BLM New Mexico State Office, *New Mexico Bureau of Land Management (BLM) Policy for Temporarily Abandoned (TA) wells and Shut-in Wells (SI)*, IM NM-2016-017 (Sept. 7, 2016). Under the BLM New Mexico state office's policy, with BLM approval, an operator can maintain the well in shut-in status for up to 2 years, after which time the operator can apply for the well to be placed in temporarily abandoned status for up to 3 years. The policy allows BLM to approve wells in shut-in status for longer than 2 years on a case-bycase basis and directs the operator to conduct production testing every 5 years.

[46] GAO-14-704G.

of the 20 BLM offices and 3 of the 10 stakeholders told us that BLM field offices have limited staff and therefore prioritize other work, such as processing drilling permits, over conducting well and bond adequacy reviews, which are used to manage potential liabilities. BLM prioritizes processing drilling permits over well and bond adequacy reviews in part because the agency is required by statute to process drilling permits within 30 days of receiving a complete application. BLM headquarters officials told us that processing permits is the agency's highest priority activity and that they ask field offices for monthly progress reports with projected goals for processing permits within the next 90 days, and compare the offices' accomplishments to agency targets. BLM headquarters officials told us that prioritizing processing permits increases the workload at the national, state-office, and field-office levels.

Officials from one BLM state office told us that other challenges to managing its potential liabilities are staffing limitations and the time it takes to conduct bond adequacy reviews. These state office officials told us that bond reviews can take a long time to complete because some bonds are associated with several hundred wells. Similarly, officials from one field office stated that conducting bond adequacy reviews was time consuming and that they had only one staff member dedicated to conducting the reviews. In 2011, we found that a lack of resources and higher agency priorities were the primary reasons for why many BLM field office officials we interviewed had not conducted well and bond adequacy reviews or did not know the number of reviews they had conducted.[47]

In addition, officials from 6 of the 20 BLM offices and 1 stakeholder told us that another challenge BLM faces in managing its potential liabilities is prioritizing funding to reclaim orphaned wells. For example, an official from one state office told us that securing funding to reclaim orphaned wells is a challenge because BLM does not set aside funding to pay for reclamation costs. BLM officials in one field office told us that they had not received funding from BLM headquarters specifically for reclamation in over 10 years, despite managing a growing number of

---

[47] GAO-11-292.

orphaned wells. An official from this field office told us that without dedicated funds from BLM headquarters for this purpose, the field office was unable to reclaim the orphaned wells. In addition, officials from another field office told us that time frames for competing and awarding contracts to perform reclamation work do not coincide with securing funding from BLM headquarters, and that funding has to be obligated by the end of the fiscal year. These officials explained that in one instance, by the time they obtained funding for well reclamation, it was too late to issue a contract for the work.

EP Act 2005 requires the establishment of a program to reclaim orphaned, abandoned, or idled oil and gas wells on federal lands. As part of this program, BLM conducts well reviews and bond adequacy reviews. As discussed above, about half of the orphaned wells BLM identified in 2009 were not reclaimed and remained orphaned in 2017, and BLM officials cited funding as the issue. The Project Management Institute, Inc. has established a standard on program management.[48] Under the standard, program resource management planning ensures that all required resources are made available for managers to enable the delivery of benefits for a program. Resource management planning involves identifying existing resources and the need for additional resources. The program manager analyzes the availability of each resource, in terms of both capacity and capability, and determines how these resources will be allocated to avoid over-commitment or inadequate support. Such planning, through a resource management plan, forecasts the expected resources across a program to allow the program manager to identify potential resource shortfalls or conflicts over the use of scarce or constrained resources. The plan is also to describe guidelines for making program resource prioritization decisions and resolving resource conflicts.

---

[48] See Project Management Institute, Inc., *The Standard for Program Management*, Fourth Edition (Newtown Square, PA: 2017). PMI® is a registered mark of the Project Management Institute, Inc. PMI is a not-for-profit association that provides global standards for, among other things, project and program management. These standards are utilized worldwide and provide guidance on how to manage various aspects of projects, programs, and portfolios.

Based on our discussions with BLM headquarters and field office officials, BLM does not have a resource management plan. For example, when we discussed resources for reclaiming orphaned wells with BLM headquarters officials, they told us that some BLM offices obtain funding from state funds established for reclaiming orphaned wells, but not all offices have been able to access such funds. If unable to secure funding from the states, offices may request funding from BLM headquarters for reclamation, and as mentioned previously, occasionally try to use unexpended funds left at the end of a fiscal year. In its comments on the draft report, Interior noted that BLM engages in annual work planning processes designed to facilitate agency resource allocation decisions. However, BLM overall does not have information on the federal resources needed to reclaim known orphaned wells. Without developing a resource management plan addressing resources needed for conducting well and bond adequacy reviews and reclaiming orphaned wells, BLM cannot have reasonable assurance that it is achieving the program's objectives.

Agency Officials and Stakeholders Identified Several Additional Challenges BLM Faces in Managing Its Potential Liabilities

Agency officials and stakeholders cited additional challenges including BLM's ability to review nationwide bonds, minimum bond amounts, and operator unresponsiveness.

Reviewing Nationwide Bonds

Officials from 10 of the 20 BLM offices told us that they encountered challenges reviewing nationwide bonds because of a lack of coordination between BLM offices. The purpose section of the bond adequacy review policy states that field offices are to review bonds to determine whether the bond amount appropriately reflects the level of potential risk posed by the operator. However, the bond adequacy review policy also states in a directive that if the bond being reviewed is a nationwide or statewide bond, field offices are only to review the wells within their field office. Officials from one field office told us that without insights into an operator's activities in the jurisdictions of other field offices, bond adequacy reviews do not cover when an operator has been cited with an *Incident of Noncompliance* or the number of inactive wells the operator may have in other jurisdictions. These field office officials said that it is important to communicate and coordinate with other field offices when there is a need to require an operator to secure a larger bond. For example, to require a well operator to increase the amount of its bond, BLM must show that the operator meets the point system's threshold in the bond adequacy review's calculation worksheet.[49] Officials in one state office told us that under a nationwide or statewide bond, an operator might not reach the agency's threshold for requiring a bond increase based on an operator's activities in the jurisdiction of one field office but may meet the threshold if BLM's bond adequacy review assessed all of the operator's operations within a state or across the nation.

Under federal standards for internal control, management should design control activities to achieve objectives and respond to risks, such as by clearly documenting internal controls, and having the documentation appear in management directives, administrative policies, or operating manuals.[50] While BLM has documented its policy, the purpose of the policy to ensure that the bond amount appropriately reflects the level of potential risk posed by the operator conflicts with a directive of the policy that offices are only to review wells within their own jurisdiction. Officials told us that BLM is currently revising the bond adequacy review policy. As the agency revises its bond adequacy review policy, BLM has the opportunity to ensure that bond adequacy reviews reflect the overall risk presented by operators. By having the policy ensure that the reviews of nationwide and statewide bonds account for overall operator risk, BLM can have better assurance that it will reduce the likelihood of using taxpayer funds to pay to reclaim orphaned wells.

---

[49] According to the bond adequacy review policy, field office officials use the point system worksheet to determine how much to adjust the bond, if necessary. For example, field office officials assign points if an operator has a history of *Incidents of Noncompliance* in the last 3 years.

[50] GAO-14-704G.

Minimum Bond Amounts

Officials from 9 of the 20 BLM offices and 1 stakeholder told us that BLM faces challenges related to federal minimum bond amounts that in their opinion are too low. For example, officials from one BLM state office expressed concerns about operators with multiple wells covered by the minimum bond amounts, which the officials believed to be inadequate to cover total potential reclamation costs. Minimum bond amounts were set in the 1950s and 1960s and have not been updated to keep up with inflation. Specifically, the $10,000 minimum for individual bonds was established in 1960, and the bond minimums for statewide bonds ($25,000) and nationwide bonds ($150,000) were established in 1951. If adjusted to 2016 dollars, these amounts would be $63,613 for an individual bond, $189,825 for a statewide bond, and $1,138,952 for a nationwide bond. According to BLM headquarters officials, the agency does not require that operators provide full liability bonds. These officials told us that they believed that most operators would not be able to remain in business if bond amounts were based on estimated total reclamation costs.

Operators' Unresponsiveness

Officials from 8 of the 20 BLM offices and 2 stakeholders told us that BLM faces challenges dealing with unresponsive operators when requiring operators to increase bond amounts or issuing *Incidents of Noncompliance*. For example, officials from one BLM state office told us that operators do not always respond to letters informing them of a requirement to secure an increase in their bond. Officials from another BLM state office told us that the agency can place operators on a noncompliance list prohibiting them from holding leases or conducting operations on federal lands. However, these officials also said that they have seen operators ask relatives to obtain leases in order to circumvent such prohibitions. Officials from one field office told us of one particular instance in which BLM had spent over 7 years attempting to enforce the requirements for reclamation activities. BLM had issued an *Incident of Noncompliance*, but the operator did not respond and instead reorganized as a separate corporate entity. Subsequently, the operator went bankrupt, requiring BLM to restart the communications process from the beginning with the newly formed entity. BLM officials told us that the agency has very little leverage when companies change their name or reorganize in an attempt to evade performing required reclamation activities. BLM headquarters officials told us that working with operators was a delicate balance, especially when oil and gas prices are down, and BLM field offices would benefit from conducting periodic operator outreach to have an open dialogue with the operators.

CONCLUSION

BLM is responsible for overseeing oil and gas development on federal lands and for balancing the sometimes competing priorities of encouraging oil and gas development, while ensuring that when wells run dry, operators return well sites to their original natural conditions. Federal laws, regulations, and BLM's own policies call for the agency to take various actions to manage its potential oil and gas well liabilities and reclaim orphaned wells. However, BLM does not systematically or comprehensively track how much the agency has spent to reclaim orphaned wells or information, such as the number of orphaned wells and inactive wells over time, necessary to determine the agency's potential liabilities. Without systematically or comprehensively tracking information on BLM's well reclamation costs and indicators of potential future costs, its ability to monitor its progress and plan for its potential liabilities associated with orphaned wells is limited.

In addition, implementation of BLM's well and bond adequacy review policies by the field offices is hampered by officials having different understandings of what constitutes a well review. This variance is because BLM's well review policy does not outline specific instructions on what actions field offices should take when conducting a well review. This situation results in inconsistent ways of conducting well reviews and annually reporting on them. Without developing and communicating specific instructions outlining what actions constitute a well review for annual-reporting purposes, BLM cannot have reasonable assurance that its field offices are conducting and reporting on well reviews in a consistent manner. Further, inaccuracies in certain AFMSS data, such as the dates that wells last changed statuses, raise questions about the quality of data BLM headquarters uses to determine the extent to which its offices are implementing the well review and bond adequacy review policies. BLM has not taken steps to improve AFMSS' data quality such as through the use of additional edit checks to prevent field offices from inputting erroneous data or having data stewards certify the quality of the data. Without taking such steps, BLM cannot have reasonable assurance that management has accurate information it needs to track whether field offices are conducting well and bond adequacy reviews as intended. In addition, BLM's approach to monitoring the implementation of its well and bond adequacy review policies is limited because the reports the agency uses to monitor implementation provide insufficient and at times conflicting information. Without taking actions to strengthen its approach to monitoring, such as collecting and analyzing data on performance indicators and ensuring the quality of those data, BLM's ability to assess the extent to which field offices are reviewing all inactive wells and determining the adequacy of all bonds will continue to be limited.

BLM officials and stakeholders identified several challenges that BLM faces in managing its potential oil and gas well liabilities, including identifying and managing certain inactive wells—specifically wells that are in shut-in status and that have the potential to become orphaned. This problem is because operators are generally not required to notify BLM when they place a well in shut-in status. Without providing greater specificity in current policy or supplemental guidance to all field offices, the federal government may face increased potential liabilities if shut-in wells become orphaned. In addition, BLM faces challenges related to limited resources and competing priorities, such as not setting aside funding to pay for reclaiming orphaned wells. Without developing a resource management plan addressing resources needed for conducting well and bond adequacy reviews and reclaiming orphaned wells, BLM cannot have reasonable assurance that it is achieving the program's objectives. BLM also faces challenges related to conducting nationwide and statewide bond adequacy reviews because the bond adequacy review policy overall contains conflicting information on how field offices are to review bonds' adequacy. BLM is currently revising the bond adequacy review policy and has an opportunity to ensure that the reviews of nationwide and statewide bonds reflect operators' overall risks.

RECOMMENDATIONS FOR EXECUTIVE ACTION

We are making the following seven recommendations to BLM:

The Director of BLM should systematically and comprehensively track the actual costs BLM incurs when reclaiming orphaned wells and the information, including the number of orphaned wells and inactive wells over time, necessary to determine the agency's potential liabilities (Recommendation 1).

The Director of BLM should develop and communicate specific instructions on what actions constitute a well review for annual-reporting purposes (Recommendation 2).

The Director of BLM should take steps to improve AFMSS data quality, for example, by conducting more edit checks and by having data stewards certify the quality of the data (Recommendation 3).

The Director of BLM should strengthen its approach to monitoring field offices' implementation of the well review and bond adequacy review policies, such as by collecting and analyzing data on performance indicators and ensuring the quality of those data (Recommendation 4).

The Director of BLM should provide greater specificity in current policy or supplemental guidance to all BLM field offices on how to identify and manage all shut-in wells (Recommendation 5).

The Director of BLM should develop a resource management plan addressing resources needed for conducting well and bond adequacy reviews and reclaiming orphaned wells (Recommendation 6).

The Director of BLM should, in revising the bond adequacy review policy, ensure that the reviews of nationwide and statewide bonds reflect the overall risk presented by operators (Recommendation 7).

AGENCY COMMENTS AND OUR EVALUATION

We provided a draft of this chapter to the Department of the Interior for review and comment. In its comments, reproduced in appendix II, Interior generally concurred with our recommendations. Interior stated that, following GAO's 2011 report on potential oil and gas well liabilities, BLM implemented comprehensive policies to better manage and minimize the risks of idle and orphaned wells on federal and Indian lands. Interior agreed that there are areas where BLM can improve the accuracy of its data and further reduce the risks associated with idle and orphaned wells.

Interior indicated that it will update and improve its existing policies and guidance consistent with the findings and recommendations in our report.

In response to our sixth recommendation—that BLM develop a resource management plan addressing resources needed for conducting well and bond adequacy reviews and reclaiming orphaned wells—Interior stated that BLM conducts annual work planning processes which facilitate decisions regarding the allocation of agency resources and requested additional information clarifying how our recommendation fits into or differs from these. We expanded our description of resource management planning and added language regarding BLM's annual work planning processes to the report. However, we were not able to review the scope or adequacy of BLM's annual work planning processes as they relate to resource planning for well and bond reviews and reclaiming orphaned wells for this chapter.

As agreed with your offices, unless you publicly announce the contents of this chapter earlier, we plan no further distribution until 30 days from the report date. At that time, we will send copies to the appropriate congressional committees, the Secretary of the Interior, and other interested parties.

Frank Rusco
Director, Natural Resources and Environment.

APPENDIX I: OBJECTIVES, SCOPE, AND METHODOLOGY

This chapter examines (1) how BLM's actual costs incurred to reclaim orphaned wells and potential oil and gas well liabilities have changed, if at all, for fiscal years 2010 through 2017; (2) the extent to which BLM has implemented its 2012 well review and 2013 bond adequacy review policies; and (3) BLM officials' and stakeholders' views on what challenges, if any, BLM faces in managing its potential liabilities.

To examine how BLM's actual reclamation costs incurred and potential oil and gas well liabilities have changed, we analyzed data in BLM's Automated Fluid Minerals Support System (AFMSS) on oil and gas wells on federal and Indian lands, including inactive wells—which represent potential liabilities.[51] We reviewed documentation provided by BLM and compared BLM's policies and procedures on recording information on actual costs incurred to reclaim orphaned wells and potential liabilities against the information and communication standard outlined in *Standards for Internal Control in the Federal Government*.[52] We selected and interviewed officials from 13 BLM field offices because, according to fiscal year 2016 data from the Department of the Interior's Office of Natural Resources Revenue (ONRR) Oil and Gas Operations Report (OGOR) data system we analyzed, these offices were responsible for about 80 percent of all oil and gas wells managed by BLM.[53] In addition, we interviewed officials from the 6 BLM state offices associated

---

[51] AFMSS is the central database that BLM uses to maintain information on oil and gas activities on federal and Indian lands. Not all data we analyzed covered the entirety of the fiscal years 2010 through 2017, due to data limitations and other issues that we describe in this chapter.

[52] GAO-14-704G.

[53] There are 127 BLM field offices, of which 33 manage oil and gas programs.

with the 13 selected field offices (see Table 1). Findings from selected offices cannot be generalized to those we did not include in our review.

However, because AFMSS does not contain information on actual costs incurred to reclaim orphaned wells, we obtained documentation of the actual reclamation costs that 13 selected BLM field offices incurred for fiscal years 2010 through July 2017. To analyze these costs, we reviewed purchase orders, invoices, and other documentation for actual reclamation work performed. We also obtained documentation, including spreadsheets with estimated potential reclamation costs that these 13 selected field offices faced as of July 2017.[54]

Table 1. Selected Bureau of Land Management (BLM) State Offices and Field Offices

BLM Office
California State Office
Bakersfield Field Office
Colorado State Office
Colorado River Valley Field Office
Montana-Dakotas State Office
North Dakota Field Office
Division of Oil and Gas—Great Falls[a]
New Mexico State Office
Carlsbad Field Office[b]
Farmington Field Office
Oklahoma Field Office
Wyoming State Office
Buffalo Field Office
Casper Field Office
Pinedale Field Office
Rawlins Field Office
Worland Field Office
Utah State Office
Vernal Field Office

Source: GAO. | GAO-18-250.

[a]Formerly Great Falls Field Office.

[b]Carlsbad includes the Hobbs field office because Carlsbad manages the Hobbs field office's wells in addition to its own.

[54] In some cases, officials from the selected field offices continued to provide us with updates on reclamation costs over the course of our review.

To assess the reasonableness of estimated reclamation costs, we reviewed estimates provided by officials from the selected field offices and compared those to historical actual costs that we previously reported in January 2010. We determined the overall estimated reclamation costs were sufficiently reasonable for providing a sense of the general magnitude of potential costs, though we did not assess the underlying inputs or assumptions used. The information we received is not generalizable to reclamation costs for other BLM offices that we did not review. We also analyzed AFMSS data on the number of wells capable of production on federal lands from fiscal years 2010 to 2016. The AFMSS database provides a snapshot of the time that the data are queried, and so does not include historical data over time. As such, to examine the number of inactive wells on federal and Indian lands and how long these have been inactive, we combined AFMSS data with data from the OGOR data system through September 2016. The Department of the Interior requires monthly OGORs from operators, which document and record the volume of oil and gas produced from wells on federal and Indian lands. From AFMSS, we identified the appropriate population of wells by selecting wells only located on federal and Indian lands, and excluded wells that were on state or private lands. Because we did not find data in AFMSS on how long a well had been in its last recorded status to be reliable, we analyzed production records from the OGOR data system. We also excluded data on wells that were in statuses in which there was no associated potential liability, such as wells pending an application for permit to drill.

For each reporting date through September 2016, we aggregated data from multiple well completions to the 10-digit unique well identifier level. We then matched the unique well identifiers in AFMSS to those listed in the OGOR data system to enumerate inactive wells by duration of inactivity. For each reporting date, we designated wells with at least one completion showing non-zero production volumes[55] or in drilling or monitoring status in the OGOR data system as active. We also designated a well as active at a certain date if AFMSS data indicated any of its

[55] We considered gas, oil, water, and injection volumes.

completions were completed on that date. Otherwise we deemed wells where all completions had zero production reporting on a date as inactive for the corresponding period. In some cases, (i) no OGOR records existed with non-zero production volumes or drilling or monitoring well status and (ii) no AFMSS well completion date was provided, and so we calculated inactivity by using the earliest record date for that well in the OGOR data set. We discussed our methodology for calculating the number of wells with BLM officials. We compared the number of inactive wells from our analysis to those reported in BLM national and state reports to identify data inconsistencies. In addition, we analyzed AFMSS reports, as of July 2017, to analyze data on the number of orphaned wells. To assess the reliability of OGOR and AFMSS data, we reviewed agency documents, met with relevant agency officials, and performed electronic testing by verifying, for example, missing or out-of-range data values. We found the data for the number of inactive wells and how long they have been inactive as well as the data for the number of wells BLM has identified as orphaned to be sufficiently reliable for our purposes.

To determine the extent to which BLM has implemented its 2012 and 2013 policies for conducting well reviews and bond adequacy reviews, we reviewed applicable laws and analyzed the well review and bond adequacy review policies.[56] We reviewed information contained in BLM's well review and bond adequacy review reports for fiscal year 2016 as well as data generated through AFMSS on bonds and wells as of October 2017. We were unable to fully assess BLM's performance against the directives in the agency's 2012 well review and 2013 bond adequacy review policies due to limited agency data and documentation as discussed in the report. Specifically, we identified data accuracy and consistency concerns with some of the data elements in the agency's well review and bond adequacy review reports as well as some AFMSS data on wells and bonds, which we discuss in this chapter. We performed electronic testing by verifying out-of-range values, such as dates of well reviews conducted that were listed as

---

[56] We examined BLM's well review policy, Instruction Memorandum 2012-181, issued on September 5, 2012; we also examined BLM's bond adequacy review policy, Instruction Memorandum No. 2013-151, issued on July 3, 2013.

being in the future. We also interviewed officials from BLM headquarters, the 13 selected field offices, and the 6 associated BLM state offices, to obtain information on the extent to which the selected offices implemented the 2012 and 2013 policy directives. We compared BLM's procedures detailing how field offices are to count or report a well review as well as procedures for maintaining data quality against the control activities standard outlined in *Standards for Internal Control in the Federal Government*.[57] We also compared BLM's procedures for monitoring implementation of policy directives against leading practices for monitoring agency policies.[58] We also reviewed documentation for a random, non-generalizable sample of 62 well reviews and 58 bond adequacy reviews, as reported by the 13 selected BLM field offices, for a total of 120 reviews. A GAO statistician selected a random sample of five well reviews for unique well numbers and five bond reviews of unique bond numbers that the 13 selected field offices had reviewed from the fiscal year 2016 well report and bond adequacy report.[59] Due to variations in field offices' reporting, some well and bond reviews from prior fiscal years were also included in the random selection. The Farmington field office also did not conduct any bond adequacy reviews in fiscal year 2016, and so we included bond reviews that the field office conducted in fiscal year 2015 in the random selection. In addition, the Pinedale and Rawlins field offices had not conducted any bond adequacy reviews in fiscal year 2016. As a result, we randomly selected additional reviews from fiscal year 2015 for those field offices. The Pinedale, Rawlins, and Colorado River Valley field offices conducted less than 5 bond reviews in each office in

---

[57] GAO-14-704G.

[58] We reviewed leading practices for monitoring government programs and projects based primarily on *Standards for Internal Control in the Federal Government*; the GPRA Modernization Act of 2010, Pub. L. No. 111-352, 124 Stat. 3866 (2011); and U.S. Office of Management and Budget, *Uniform Administrative Requirements, Cost Principles, and Audit Requirements for Federal Awards*, 2 C.F.R. part 200.

[59] We originally developed a data collection instrument to use when reviewing physical files, but following initial testing at two field offices in Wyoming, determined that the physical files for the selected well reviews did not have significant documentation. The team instead selected the random, non-generalizable sample of five well reviews and five bond reviews conducted by each field office in fiscal year 2016, and requested electronic supporting documentation on those reviews based on the unique well number and unique bond number attached to those reviews.

that fiscal year, so we selected and reviewed documentation in support of only those reviews they had conducted. We assessed the documentation to determine whether or not field offices conducted reviews and complied with selected directives of the well review and bond adequacy review policies. Information from our documentation reviews is not generalizable to all BLM field offices but provides illustrative examples of the information contained in BLM well and bond adequacy reviews.

To examine BLM officials' and stakeholders' views on what challenges, if any, BLM faces in managing its potential oil and gas well liabilities, we conducted semi-structured interviews with officials from BLM headquarters, the 13 selected BLM field offices, and the 6 BLM state offices associated with these 13 field offices. In addition, we interviewed or obtained written responses from a standard set of questions from 8 representatives of stakeholder organizations. These representatives were knowledgeable about BLM's oil and gas well management, and included academic, environmental, industry, and state organizations (see Table 2). In addition, we spoke with knowledgeable officials from the Department of the Interior's Office of Natural Resources Revenue (ONRR) and the Department of the Interior's Office of Indian Energy and Economic Development, Division of Energy and Mineral Development.[60] To identify knowledgeable stakeholders, we conducted a literature search, reviewed previous GAO reports, and obtained recommendations from BLM officials and stakeholders using a snowball technique in which an initial group of BLM officials and stakeholders we interviewed identified additional contacts to interview. From this list, we selected stakeholders who could provide a range of viewpoints. We generally asked the same questions during each interview but also discussed individual stakeholders' perspectives, as appropriate. In our interviews, we asked officials and stakeholders what challenges, if any, BLM offices face in managing their potential oil and gas well liability. We also asked what challenges, if any, BLM offices face in conducting well reviews and bond adequacy reviews. To identify the challenges identified most often in the interviews, two

---

[60] For purposes of this chapter, we refer to these 10 stakeholders and Department of the Interior officials collectively as stakeholders.

analysts developed categories of challenges identified by BLM offices and stakeholders, and each analyst independently determined whether each BLM office and stakeholder had identified challenges that fit into these categories. The two analysts discussed and resolved any differences in their coding. The views of the BLM officials, stakeholders, and other agency personnel we interviewed are not generalizable to BLM officials, similar stakeholders, and other agency personnel who we did not interview. Lastly, we compared how BLM identified and managed certain inactive wells, as well as how it managed nationwide and statewide bonds, against the control activities standard outlined in *Standards for Internal Control in the Federal Government*[61] and BLM's resource management practices against certain requirements in the Energy Policy Act of 2005 (EPAct 2005) and leading practices by the Project Management Institute in *The Standard for Program Management*.[62]

Table 2. Stakeholders and Other Agencies GAO Interviewed

Stakeholders and other agencies within the Department of the Interior	Category
Colorado Oil and Gas Conservation Commission	State
Department of the Interior, Office of Indian Energy and Economic Development, Division of Energy and Mineral Development	Other
Department of the Interior, Office of Natural Resources Revenue	Other
Environmental Defense Fund	Environmental
Interstate Oil and Gas Compact Commission	State
Petroleum Association of Wyoming[a]	Industry
Professor Lucas Davis, University of California-Berkeley	Academia
Resources for the Future	Environmental
Western Organization of Resource Councils	Environmental
Wyoming Oil and Gas Conservation Commission	State

Source: GAO. | GAO-18-250.

[a]Representatives from the Petroleum Association of Wyoming declined to meet because they were not knowledgeable about all our questions. They instead provided written comments to selected questions they felt they could answer. We also contacted the Independent Petroleum Association of America and Western Energy Alliance, but did not interview representatives from either organization because representatives told us that they were not knowledgeable about BLM's management of orphaned and inactive wells.

---

[61] GAO-14-704G.

[62] Project Management Institute, Inc., *The Standard for Program Management*, Fourth Edition (Newtown Square, PA: 2017). PMI® is a registered mark of the Project Management Institute, Inc.

We conducted this performance audit from November 2016 to May 2018, in accordance with generally accepted government auditing standards. Those standards require that we plan and perform the audit to obtain sufficient, appropriate evidence to provide a reasonable basis for our findings and conclusions based on our audit objectives. We believe that the evidence obtained provides a reasonable basis for our findings and conclusions based on our audit objectives.

APPENDIX II: COMMENTS FROM THE DEPARTMENT OF THE INTERIOR

United States Department of the Interior
OFFICE OF THE SECRETARY
Washington, DC 20240

APR 3 0 2018

Mr. Frank Rusco
Director
Natural Resources and Environment
U.S. Government Accountability Office
441 G Street NW
Washington, DC 20548

Dear Mr. Rusco:

Thank you for giving the Department of the Interior (Department) the opportunity to review and comment on the draft Government Accountability Office (GAO) report entitled, *Federal Oil and Gas Wells: Bureau of Land Management Needs Better Data and Oversight of Its Potential Liabilities* (GAO-18-250).

The Department is committed to sustainably developing oil and gas resources on Federal and Indian lands in a manner that recognizes the need for domestic sources of minerals resources, while minimizing potential liabilities to the United States Government. Following the 2011 GAO Report *Oil and Gas Bonds: BLM needs a Comprehensive Strategy to Better Manage Potential Oil and Gas Liability*, the Bureau of Land Management (BLM) implemented comprehensive policies to better manage and minimize the risks of idle and orphan wells on Federal and Indian lands. While these policies have provided an appropriate framework for managing potential liabilities from oil and gas wells, we agree that there are areas where the BLM can improve the accuracy of its data on non-operational wells and further reduce the risks associated with idle and orphan oil and gas wells on Federal and Indian lands. The BLM will update and improve its existing policies and guidance, while pursuing technological enhancements consistent with the audit findings and recommendations in this draft report.

The GAO issued seven recommendations in response to its overall findings. The Department generally agrees with the findings and concurs with the recommendations. However, we request clarification on Recommendation 6 as it relates to the term "resource management plan."

Below is a summary of the BLM's response to the recommendations and the actions planned to implement the recommendations.

Recommendation 1: The Director of BLM should systematically and comprehensively track the actual costs BLM incurs when reclaiming orphaned wells and information necessary to determine the agency's potential liabilities, including the number of orphaned wells and inactive wells over time.

Response: The BLM concurs with this recommendation. The BLM will expand existing reporting capabilities to provide the ability to track orphaned and inactive well records over time. The BLM will also develop a new funding code to track the costs incurred when reclaiming orphaned wells. The BLM will issue written guidance to state and field offices in order to implement the bureau-wide use of the code(s) to track the costs incurred reclaiming orphaned wells.

Recommendation 2: The Director of BLM should develop and communicate specific instructions on what actions constitute a well review for annual reporting purposes.

Response: The BLM concurs with this recommendation. The BLM will provide additional guidance by updating IM 2012-181 *Idle Well Review and Data Entry into the Automated Fluid Minerals Support System (AFMSS)*, to further bolster and communicate appropriate actions for well reviews. While the BLM believes that its existing policies provide considerable guidance to state and field offices on how to conduct and report non-operational well reviews, the BLM also believes that it would be beneficial to develop and issue written policy updating and replacing IM 2012-181. This new and updated guidance will provide state and field offices additional clarification as to what actions constitute a well review. The BLM is also currently pursuing system enhancements in the updated AFMSS 2 database that will improve the quality and reporting capabilities of its non-operational well data, which will also be incorporated into the written policy that updates and replaces IM 2012-181.

Recommendation 3: The Director of BLM should take steps to improve AFMSS data quality by, for example, conducting more edit checks, and having data stewards certify the quality of the data.

Response: The BLM concurs with this recommendation. The BLM currently performs AFMSS 1 and AFMSS 2 data quality checks on an ad hoc basis. The BLM will take additional steps to further improve AFMSS 1 and AFMSS 2 data quality by developing guidance on the frequency and standard procedures for data validation review and certification which will also be part of the IM-2012-181 update.

Recommendation 4: The Director of BLM should strengthen its approach to monitoring field offices' implementation of the well review and bond adequacy review policies, such as by collecting and analyzing performance indicators and ensuring the quality of those data.

Response: The BLM concurs with this recommendation. The BLM will update its current data review processes by developing guidance on the frequency and standard procedures for data validation reviews and provide more detailed instructions on documenting idle well and bond adequacy reviews. Those actions, combined with AFMSS 2 system enhancements that will improve the data quality and reporting capabilities, will strengthen the BLM's monitoring of the implementation of its well review and bond adequacy review policies.

Recommendation 5: The Director of BLM should provide greater specificity in current policy or supplemental guidance to all BLM field offices on how to identify and manage all shut-in wells.

Response: The BLM concurs with this recommendation. As previously stated in response to Recommendation 2, IM 2012-181 already provides BLM state and field offices with guidance on reviewing and reporting reviews of wells, including those in shut-in (SI) status. However, the BLM will issue new guidance on conducting, tracking, and reporting SI well reviews. As part of the ongoing AFMSS 2 database enhancements, the BLM will improve the quality and reporting capabilities of its SI well data, which will also be incorporated in the new guidance replacing IM 2012-181.

Recommendation 6: The Director of BLM should develop a resource management plan addressing resources needed for conducting well and bond adequacy reviews and reclaiming orphaned wells.

Response: The BLM conducts annual work planning processes which facilitates the decisions regarding the allocation of agency resources. The BLM is requesting additional information regarding the "resource management plan" envisioned and recommended by the GAO to clarify how the suggested plan fits into or differs from the established Bureau and Department planning processes.

Recommendation 7: The Director of BLM should, in revising the bond adequacy review policy, ensure that the reviews of nationwide and statewide bonds reflect the overall risk presented by operators.

Response: The BLM concurs with this recommendation. The BLM will issue new guidance replacing the expired IM 2013-151 entitled, "Oil and Gas Bond Adequacy Reviews." This new guidance will ensure that the reviews account for the overall operator risk by clarifying the circumstances which warrant the adjustment of bond amounts.

Sincerely,

Joseph R. Balash
Assistant Secretary
Land and Minerals Management

In: Focus on Oil and Gas
Editor: Olivia B. Wilder
ISBN: 978-1-53616-325-4

Chapter 2

OIL AND GAS LEASE MANAGEMENT: BLM COULD IMPROVE OVERSIGHT OF LEASE SUSPENSIONS WITH BETTER DATA AND MONITORING PROCEDURES*

United States Government Accountability Office

ABBREVIATIONS

BLM	Bureau of Land Management
LR2000	Legacy Rehost 2000 System

* This is an edited, reformatted and augmented version of the United States Government Accountability Office Report to Congressional Requesters, Publication No. GAO-18-411, dated June 2018.

Why GAO Did This Study

Oil and gas leases on federal lands generate billions of dollars in rents and royalty payments for the federal government each year, but these revenues can be reduced if leases are suspended (i.e., placed on hold). Questions have been raised about whether some suspensions, particularly those in effect for more than 10 years, may hinder oil and gas production or adversely affect the use of federal lands for other purposes, such as recreation.

GAO was asked to review oil and gas lease suspensions on federal lands managed by BLM. This chapter examines, among other things, (1) the extent of and reasons for such suspensions and (2) the approach BLM uses to monitor the status of lease suspensions. GAO analyzed all data on suspensions in a BLM database and the official lease files for a nongeneralizable sample of 48 leases recorded as suspended in, Montana and Wyoming, which GAO selected based in part on the large number of suspensions these states had. GAO also reviewed BLM documents and interviewed BLM officials.

What GAO Recommends

To better ensure that federal lands are not being inappropriately kept from development, GAO is making four recommendations, including that BLM record the reasons for lease suspensions in its database and develop procedures for monitoring suspensions. Interior concurred with GAO's recommendations.

What GAO Found

According to data at the end of fiscal year 2016 from the Bureau of Land Management (BLM), a small portion of oil and gas leases were suspended for various lengths of time (as shown below), but the reasons for the suspensions were difficult to determine. During a suspension, the government generally does not collect revenues from the lease. Determining the reasons for suspensions is difficult, in part because BLM does not require the inclusion of this information in its database. To obtain this information, BLM officials would have to review the official lease files, of which many are in hard copy. Under *Standards for Internal Control in the Federal Government*, management should use quality information to achieve the entity's objectives. BLM field officials GAO interviewed said that additional, more detailed information in the database on reasons for suspensions would be helpful in tracking lease suspensions. By including a data field in the database to record the reasons for suspensions, BLM could better ensure that federal lands are not being inappropriately kept from development—potentially foregoing revenue—or from other valuable uses of public lands.

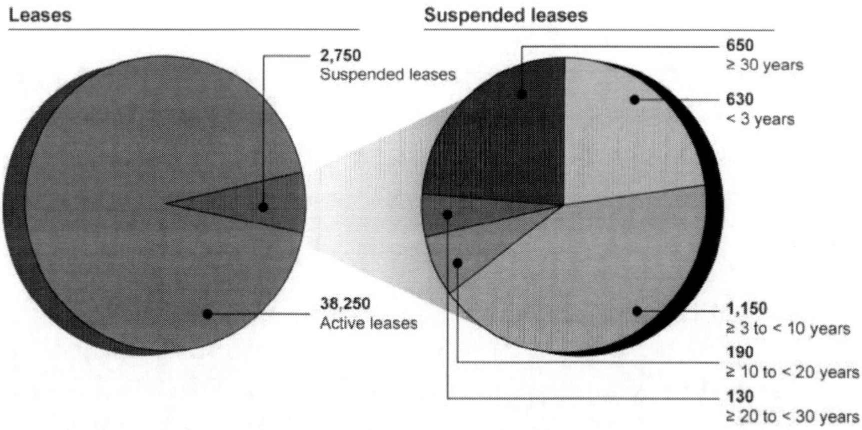

Source: GAO analysis of Bureau of Land management data. | GAO-18-411.
Note: Numbers represented in the figure have been rounded.

Active Oil and Gas Leases and leases recorded as in suspension by length of suspension, as of September 30, 2016

The approach BLM uses to monitor lease suspensions does not ensure consistent and effective oversight because BLM does not have procedures in place for monitoring. BLM state offices generally delegate responsibility for monitoring lease suspensions to their field offices. Officials from 12 selected field offices in two states with relatively large numbers of lease suspensions reported various frequencies in their monitoring of suspensions, ranging from every few months to rarely or not at all. In the absence of BLM monitoring procedures, field officials have discretion in how and when to monitor. By developing procedures for monitoring lease suspensions, including when to conduct monitoring efforts, BLM could better ensure that lease suspensions in effect are warranted.

June 4, 2018

The Honorable Raúl Grijalva
Ranking Member
Committee on Natural Resources
House of Representatives

The Honorable Alan Lowenthal
Ranking Member
Subcommittee on Energy and Mineral Resources
Committee on Natural Resources
House of Representatives

The Honorable Jared Polis
House of Representatives

Oil and gas leases on onshore federal lands managed by the Department of the Interior's Bureau of Land Management (BLM) generate billions of dollars in rents and royalty payments for the federal government each year. According to BLM's database for mineral and land use—the Legacy Rehost 2000 System (LR2000)—approximately 41,000 oil and gas leases cover 28 million subsurface acres. Under the terms of these leases, operators have 10 years to begin drilling for oil and gas unless BLM suspends their leases.[1]

[1] Lease terms may be extended if, for example, oil or gas is being produced in paying quantities. *See* 30 U.S.C. § 226(e).

BLM may suspend leases, for example, when an environmental review is necessary to avoid harm to endangered species on the leased lands or during periods of extraordinary weather conditions, such as hurricanes or other natural catastrophes. Suspensions typically remain in effect until the conditions that led to the suspensions no longer apply. There is no limit on how long a lease can be suspended. During a suspension of operations and production, the government does not collect rents or royalty payments and the time remaining to begin drilling is reserved so that the operator is not penalized for the time the lease is in suspension.

BLM is responsible for implementing the Federal Land Policy and Management Act of 1976, as amended, which directs Interior to manage federal lands for multiple uses, such as recreation and resource extraction, while also taking any action required to prevent "unnecessary or undue degradation" of these lands.[2] Members of Congress and others have raised questions about whether some suspensions, particularly those that have been in effect for more than 10 years, may hinder oil and gas production or adversely affect the use of federal lands for other purposes, such as recreation.

You asked us to review oil and gas lease suspensions on lands that BLM manages. This chapter examines (1) the process BLM uses to determine whether to suspend oil and gas leases; (2) the extent of oil and gas lease suspensions and the reasons for the suspensions of selected leases; and (3) the approach that BLM uses to monitor the status of lease suspensions and the extent to which this approach provides for oversight of such lease suspensions.

To examine the process BLM uses to determine whether to suspend oil and gas leases, we reviewed applicable laws, agency documents, and the criteria used by BLM officials when considering lease suspensions. We also interviewed BLM officials about how they use these criteria. To examine the extent of oil and gas lease suspensions and the reasons for the suspensions, we analyzed data on all lease suspensions recorded in BLM's

---

[2] 43 U.S.C. § 1732.

LR2000 database as of September 30, 2016.[3] We took a number of steps to assess the reliability of these data—such as checking the extent to which the data were complete and within expected ranges and interviewing BLM officials about how the data were collected—and determined that the data were sufficiently reliable to give a high-level summary of suspension information, such as the number of leases in suspension and the lengths of the suspensions (see app. I for additional detail).

We also analyzed information contained in the official lease files, maintained by BLM state offices, for a nongeneralizable sample of leases in suspension in Montana and Wyoming to identify the reasons behind the suspensions as well as to assess their status. We chose these states because they were among the states with the largest numbers of recorded lease suspensions. We also selected Montana because their files were electronically maintained and easily accessible and Wyoming because it had some leases recorded as suspended for the longest period of time of any state as of September 30, 2016. We selected 48 suspended leases in these two states and limited the extent to which we selected multiple leases that were suspended at the same time for the same reason. For Montana leases, we found that there were only 12 suspension dates. Therefore, we randomly selected a single lease from each date. For Wyoming, we selected leases based on a combination of suspension date, similarity of lease numbers, and field office of jurisdiction. In addition, to review suspensions that had been in effect for a relatively long period of time, we selected suspensions that had been in effect for 20 or more years. The results of our review of selected leases are not generalizable to other BLM leases in suspension but provide examples of the types of reasons given for suspensions.

To verify the status of each selected lease, we compared information in LR2000 and the official lease file with information in the Office of Natural Resources Revenue's database.[4] We also interviewed officials from the BLM state and field offices in our sample about specific lease files to learn

---

[3] LR2000 contains information on all mineral and land assets managed by BLM, including oil and gas leases.

[4] The Office of Natural Resources Revenue, within Interior, is responsible for collecting rental and royalty payments associated with oil and gas leases.

more about the status of specific lease suspensions and the reasons why these suspensions were still in effect. We compared how BLM maintains and verifies its lease suspension information with *Standards for Internal Control in the Federal Government* for information and communication.[5] To examine the approach BLM uses to monitor the status of lease suspensions and the extent to which this approach provides for oversight of such lease suspensions, we reviewed agency databases and documents, including information in LR2000 and selected lease files from the Montana and Wyoming state offices. These two state offices comprised a total of 12 field offices that were responsible for monitoring the 48 suspended leases in our sample.[6] We also interviewed BLM officials from BLM headquarters and those state and field offices to learn about how, if at all, they monitor lease suspensions or oversee such monitoring. We compared their actions with agency guidance, federal regulations, and federal standards for internal control for control activities.[7]

We conducted this performance audit from September 2016 to June 2018 in accordance with generally accepted government auditing standards. Those standards require that we plan and perform the audit to obtain sufficient, appropriate evidence to provide a reasonable basis for our findings and conclusions based on our audit objectives. We believe that the evidence obtained provides a reasonable basis for our findings and conclusions based on our audit objectives.

BACKGROUND

This section provides information on oil and gas leasing and development on federally managed lands, lease revenues, lease suspensions, and BLM's LR2000 database.

---

[5] GAO, *Standards for Internal Control in the Federal Government*, GAO-14-704G (Washington, D.C.: September 2014).

[6] Montana had 2 field offices and Wyoming had 10 field offices with leases included in our review of selected lease files. Montana had 8 additional field offices that did not have leases in our sample.

[7] GAO-14-704G.

Oil and Gas Leasing and Development on Federally Managed Lands

BLM is responsible for managing approximately 700 million acres of subsurface mineral estate throughout the country, including the acreage it leases to operators for oil and gas development.[8] At the end of fiscal year 2016, about 41,000 oil and gas leases accounted for approximately 28.2 million acres in 32 states, according to BLM data (see app. II for additional details).

The Federal Land Policy and Management Act of 1976, as amended, requires the Secretary of the Interior to develop land use plans for public lands.[9] These plans identify federal lands and mineral resources that will be available for oil and gas leasing and development and other activities. The act requires the plans to be revised as appropriate, and BLM generally evaluates plans for potential revisions at least every 5 years.[10]

As part of developing or revising land use plans, BLM is required under the National Environmental Policy Act of 1969, as amended, to evaluate likely environmental effects of any decisions in the plan, such as selecting areas for oil and gas development.[11] Generally, Interior prepares an environmental impact statement—a detailed statement of the likely environmental effects of the proposed action—in preparing land use plans.[12] BLM officials said the agency uses the land use plans and environmental impact statements to (1) help develop "reasonably foreseeable development scenarios" to estimate outcomes, such as the

---

[8] The Mineral Leasing Act of 1920, as amended, gives Interior responsibility for oil and gas leasing on federal and private lands where the federal government has retained mineral rights. 30 U.S.C. § 226. The Mineral Leasing Act for Acquired Lands of 1947 (30 U.S.C. § 351 et seq.) extended that authority to lands acquired by the federal government.

[9] 43 U.S.C. § 1712.

[10] 10Revisions to land use plans are necessary if monitoring and evaluation findings, new data, new or revised policy, or changes in circumstances indicate that an entire plan or a major portion of a plan no longer serves as a useful guide for resource management. Bureau of Land Management, Department of the Interior, Land Use Planning Handbook, H-1601-1, p. 46 (Washington, D.C.: 2005).

[11] 42 U.S.C. § 4332.

[12] BLM may also use an environmental assessment—which is a more concise analysis developed if the environmental effect of the proposed action is unknown—to determine whether the action is likely to significantly affect the environment.

number of wells and likely surface disturbance that may occur under the land use plan; (2) identify lands open and closed to leasing; (3) identify resource-protection measures, such as lease stipulations and environmental best management practices; and (4) establish monitoring protocols. With a completed land use plan and its associated environmental impact statement, BLM can offer for lease the mineral rights identified in the plan.

The parcels of land that BLM offers for potential leasing and development are nominated by industry and the public or identified by BLM. BLM offers leases through a competitive bidding process and requires a uniform national minimum bid of $2 per acre, due as a one-time payment when a bidder is awarded the lease.[13] If BLM receives any bids on an offered lease, the lease is awarded to the bidder with the highest bid.[14] Since 1992, BLM has offered leases with a 10-year primary term—the initial period of time prescribed in a lease to begin oil and gas development.[15] Operators generally begin oil and gas exploration on leased lands by analyzing available geologic and seismic information and other testing to determine if economically viable oil and gas reservoirs exist. If the findings are positive, the operators may begin efforts to prepare for development, such as completing the environmental studies required to apply for permits to begin lease development activities. For example, operators holding leases for oil and gas development must submit a drilling permit application to BLM and obtain approval before preparing the land and drilling new oil or gas wells.[16] After receiving a permit application, BLM generally communicates with operators until they provide all of the required documents, including necessary environmental information or

---

[13] The Secretary of the Interior has the authority to raise the uniform national minimum acceptable bid for all leases but has not done so. *See* 30 U.S.C. § 226 (b)(1)(B).

[14] Leases that do not receive any bids in the initial offer are offered noncompetitively the day after and remain available for leasing for a period of 2 years after the competitive lease sale.

[15] The Energy Policy Act of 1992 required BLM to offer all competitive and noncompetitive leases with 10-year primary terms. Prior to 1992, BLM offered primary lease terms of 5 years for competitively sold leases and 10 years for leases issued noncompetitively.

[16] A complete drilling permit application, formally called an application for permit to drill, must include, among other things, a Surface Use Plan of Operations with the operator's plan for reclaiming disturbed lands during production and upon final abandonment. BLM may approve the application as submitted or, according to BLM officials, approve it with certain conditions that ensure environmental protection, safety, or conservation of mineral resources, and the approval may be based on environmental best management practices.

studies.[17] The Energy Policy Act of 2005 requires BLM to approve or defer permit applications within 30 days of submission by the operator.[18] After such applications are approved, operators may begin development activities, including building roads to the well site, constructing platforms, drilling wells, and constructing additional pipeline transportation necessary to transport the oil and gas to market.

BLM has the authority to inspect federal oil and gas sites, including well pads and production facilities,[19] under the Federal Oil and Gas Royalty Management Act of 1982, as amended.[20] According to the agency's handbook for its inspection and enforcement program, BLM must ensure that oil and gas operations on federal lands are prudently conducted in a manner that ensures protection of the surface and subsurface environment.

Lease Revenues

For issued leases, the operator pays a fixed amount of rent each year until the lease begins producing or expires. Under the Mineral Leasing Act of 1920, as amended, once a federal lease begins producing, the operator pays royalties on the oil and gas it produces in lieu of paying rent.[21] The act sets the royalty rate for competitive leases at not less than 12.5 percent of the amount or value of production.[22] A producing lease remains in effect so long as the operator continues to produce oil and gas in paying quantities.[23]

---

[17] Prior to leasing lands, BLM conducts broad environmental assessments. However, before drilling operations can begin, the operator must submit, with their application for permit to drill, an assessment of the environmental impact of their specific lease development plans.

[18] Pub. L. No. 109-58, § 366, 119 Stat. 594, 726 (2005), codified at 30 U.S.C. § 226(p)(2).

[19] Prior to drilling, an operator prepares the area of land where drilling will take place, referred to as a well pad. In some cases, multiple wells will be located on a single well pad.

[20] Pub. L. No. 97–451, tit. 1, § 108, 96 Stat. 2447, 2453 (1983), codified as amended at 30 U.S.C. § 1718.

[21] 30 U.S.C. § 226(d); 43 C.F.R. § 3103.2-2.

[22] 30 U.S.C. § 226(b)(1).

[23] 30 U.S.C. § 226(b)(3)(D); 43 C.F.R. § 3107.2-1.

The Office of Natural Resources Revenue, within Interior, is responsible for managing and collecting revenues from operators that produce or extract resources from federal leases. In fiscal year 2016, approximately 164 million barrels of oil and 3.25 trillion cubic feet of gas were produced on federal lands, according to agency data. According to Office of Natural Resources Revenue data, in fiscal year 2016, the federal government collected approximately $1.6 billion in gross revenue from the production of these resources on federal land. The majority of this revenue—nearly $1.5 billion, or 91 percent—came from royalties. The remaining revenue came from bids made on new leases—more than $120 million—and rent for existing leases—more than $20 million.

Lease Suspensions

According to agency guidance, specifically the *Suspensions of Operations and/or Production Manual*, BLM generally uses two types of suspensions for oil and gas leases: (1) suspension of operations or (2) suspension of operations and production.[24] A suspension of operations halts the operations associated with a particular lease, such as drilling or developing a well pad and roads. A suspension of operations and production—the most common type of suspension, according to BLM officials—is broader because it halts both operations and any production of oil and gas. BLM's guidance also states that a suspension of operations may be granted in cases in which the operator is prevented from operating or producing on the lease for reasons beyond the operator's control, and a suspension of operations and production may be granted only in the interest of the conservation of natural resources.[25] During either type of

[24] There is a third type: suspension of production; however, this type of suspension is seldom used, according to BLM officials. *See* 43 C.F.R. § 3103.4-4.

[25] The Mineral Leasing Act of 1920, as amended, defines the types of suspensions and the criteria that need to be met for each. Under Section 17, BLM can suspend either operations or production, while suspensions under Section 39 suspend both operations and production. Section 17 suspensions halt the term of the lease but rental and, if applicable, royalty payments continue. Section 39 suspensions of operations and production halt the term of the lease as well as any rent and royalty payments during the duration of the suspension.

suspension, the time remaining in the primary term of the lease is reserved until the suspension is terminated, so that the operator is not penalized for the time the lease is in suspension.

According to BLM officials, lease suspensions typically are initiated by the operator but may also be initiated by BLM. BLM guidance states that before an operator can request a suspension, the operator must first demonstrate being hampered in performing some operation or activity on the lease. The operator must submit thorough documentation of the reason for requesting a suspension and should include evidence that activity has been attempted on the lease, such as filing an application for a drilling permit, and that the activity has been prevented by actions beyond the operator's control. For BLM's part, according to BLM's guidance, requests filed less than 30 days prior to the expiration of the lease are considered late and should normally be denied. If a request is filed in a timely manner, BLM is to assess the request and, if the reasons for the request are acceptable and justify a suspension, BLM should approve the request, according to BLM guidance. The state director at each BLM state office is responsible for reviewing and approving requests for lease suspensions; however, BLM's guidance encourages the delegation of this responsibility to the field manager at the field office with jurisdiction over the lease. According to BLM officials, BLM state offices generally delegate responsibility for monitoring lease suspensions to their field offices.

BLM's LR2000 Database and Other Interior Databases

According to BLM officials, LR2000 is a national database that provides internal and external users with access to, among other things, land and mineral use authorizations for oil, gas, and other mineral development; land titles; and other data extracted from case files that support BLM land, mineral, and resources programs. LR2000 contains

---

According to BLM guidance, suspension of production is unnecessary in most cases since the ability of the lease to produce oil or gas will generally prevent the lease from expiring.

information on approximately 6 million land and mineral case files. BLM designed the database for use by the oil and gas industry, mining industry, land and mineral title companies, utilities, state and local governments, interest groups, and members of the public that need access to BLM land and mineral case files.

The agency has conducted a series of reviews of LR2000 over the last 5 years in an attempt to improve the accuracy of the data in the system, according to BLM officials we interviewed.[26] In particular, the officials informed us that they created a tool, known as Data Flux, to improve the accuracy of the data, and that the tool has helped identify numerous data errors.[27] BLM officials told us that each spring a report is generated using Data Flux that highlights the errors found in LR2000, and BLM state offices are responsible for taking action to address the identified errors for their respective states. These officials also told us that BLM plans to either significantly update or replace LR2000 but has not set a definitive date for doing so.

Interior and BLM manage several other databases that contain information about the development and production of oil and gas on federal lands. In prior work, we found weaknesses in how Interior tracks and uses some information in its data systems. Specifically, in July 2010, we reported that BLM's publicly available data related to protests, or challenges, to lease sales were incomplete or inconsistent, and we recommended that Interior determine and implement an agency-wide approach for collecting protest information that is complete, consistent, and available to the public.[28] BLM agreed with the recommendation and issued guidance to standardize data collection. In addition, we found in July 2016 that Interior could improve the data it collects to help track progress toward its goal of reducing methane

[26] These reviews were intended to improve the quality of the data in LR2000 but did not necessarily address information contained in official lease files.

[27] According to agency officials, the Data Flux tool uses a set of business rules to examine data in LR2000 in order to identify data that do not match the prescribed format, such as numerical fields that contain letters.

[28] GAO, *Onshore Oil and Gas: BLM's Management of Public Protests to Its Lease Sales Needs Improvement*, GAO-10-670 (Washington, D.C.: July 30, 2010).

emissions from oil and gas operations.[29] We made four recommendations to improve BLM's reporting of emissions data. The agency generally concurred with all of the recommendations and has implemented two of them. Further, in April 2017, we found that BLM field offices had not effectively used data collected during environmental inspections, which could have enhanced BLM's ability to assess and mitigate environmental impacts.[30] We recommended that BLM develop guidance and consistently track inspections data, among other things. BLM generally concurred with these recommendations.

BLM Uses a Multistep Process in Determining Whether to Suspend Leases

BLM uses a multistep process to determine whether to suspend oil and gas leases, and this process, according to BLM guidance and officials, typically begins with an operator submitting a suspension request to the appropriate BLM field office.[31] Once the request is received, the cognizant BLM field official—usually a petroleum engineer at the field office—reviews it for completeness and whether the reasons cited meet the suspension criteria established in federal regulations and BLM's Suspensions of Operations and/or Production Manual. These criteria require that lease suspensions be approved only in the interest of the conservation of natural resources or for circumstances beyond the operator's control. Officials we interviewed stated that field officials generally have broad discretion in how to apply suspension criteria when considering a request.

---

[29] GAO, *Oil and Gas: Interior Could Do More to Account for and Manage Natural Gas Emissions*, GAO-16-607 (Washington, D.C.: July 7, 2016).

[30] GAO, *Oil and Gas Development: Improved Collection and Use of Data Could Enhance BLM's Ability to Assess and Mitigate Environment Impacts*, GAO-17-307 (Washington, D.C.: Apr. 25, 2017).

[31] In some cases, suspensions can be initiated, or directed, by BLM in response to unique circumstances, such as a court order.

See figure 1, below, for examples of circumstances for which suspensions can be issued.

Suspension of operations and production	Suspension of operations
The Bureau of Land Management (BLM) orders a suspension of all operational activities on a lease to protect natural resources (e.g., delay oil and gas drilling to allow extraction of coal).	Other federal or state agencies' actions prevent commencement or continuation of operations.
BLM denies proposal to operate for reasons of conservation of natural resources.	Extraordinary weather conditions—that is, conditions that are not reasonably expected or that are more severe than reasonably expected for the location and time of year—or other catastrophe prevent road construction or drilling.
BLM or other surface managing agency initiates environmental studies (e.g., environmental assessments, environmental impact statements, or resource management plans) that prohibit the use of the lease.	Extraordinary weather conditions prevent surveys of cultural resources or threatened and endangered species.
BLM or other surface-managing agency needs more time to arrive at the decision on a proposal.	The operator pursues litigation over title to lease or surface access.
Environmental litigation is ongoing related to issuance of leases or BLM lease management, and meanwhile no proposal to operate can be approved.	BLM denies proposal to operate for reasons other than for conservation of natural resources.

Source: BLM's Suspensions of Operations and/or Production Manual. | GAO-18-411.

Figure 1. Circumstances That May Warrant Suspension of Oil and Gas Leases.

According to BLM officials, if the field office recommends approving the operator's request for suspension, the field office is to forward the request to the appropriate BLM state office for final review, as shown in figure 2 below. In cases in which the state office agrees with the field office's recommendation, the state office is to issue a decision letter to the operator noting the changes to the terms and conditions of the lease. A copy of the letter is also to be sent to the Office of Natural Resources Revenue, if necessary, requesting deferment of rent and royalty payments while the lease is suspended.

Conversely, if the field office recommends that the suspension request be denied, the field office is to inform the operator in writing, BLM officials said. According to agency guidance, the operator can appeal the field office's recommendation to the state office director within 20 days after receiving the notification. The state director then has 10 days to render a decision.[32] If the state director denies the request for suspension, the operator can challenge the decision at the Interior Board of Land

---

[32] The state director's final decision is to be issued within 10 days of the receipt of a complete request for administrative review or within 10 days after an oral presentation has been made.

Appeals.[33] After the board's decision, the operator may make additional appeals in federal court. In cases in which a decision is overturned, the state office is to issue a decision letter to the operator that highlights changes in the lease's terms and conditions. The state office is to record the new terms and conditions in LR2000, notify the Office of Natural Resources Revenue of any rental or royalty payments that are to be deferred, and update the official lease file in the state office. Other affected parties (i.e., any party who is adversely affected by a decision) can also appeal a suspension decision, according to BLM officials.

Agency officials stated that BLM field offices are primarily responsible for monitoring the status of lease suspensions they issue to ensure that the conditions for granting the suspension still exist.[34] If the conditions have changed, the field office is to recommend that the lease suspension be terminated and notify the operator. The state office is to terminate the suspension and send a letter to the operator with the updated lease terms and conditions, which should extend the original lease expiration date to reflect the length of the suspension.[35] BLM guidance states that the state office also is to send a copy of the suspension termination letter to the Office of Natural Resources Revenue to alert that office that any rental and royalty payments on hold for the lease should resume. The state office is then responsible for updating LR2000 and the official lease file regarding any new lease terms and conditions, according to BLM officials.

---

[33] The Interior Board of Land Appeals is an appellate review body that exercises the delegated authority of the Secretary of the Interior to issue final decisions for the Department of the Interior. Its administrative judges decide appeals from bureau decisions relating to the use and disposition of public lands and their resources, mineral resources on the Outer Continental Shelf, and the conduct of surface coal mining operations under the Surface Mining Control and Reclamation Act. Located within the Department's Office of Hearings and Appeals, the board is separate and independent from the bureaus and offices whose decisions it reviews.

[34] BLM officials we interviewed stated that most lease suspensions are approved by field offices and the field offices would be responsible monitoring these suspensions. Suspensions issued by the state office would be monitored by the state office.

[35] For example, if a suspension was in effect for 3 years, the primary lease term would be extended by 3 years.

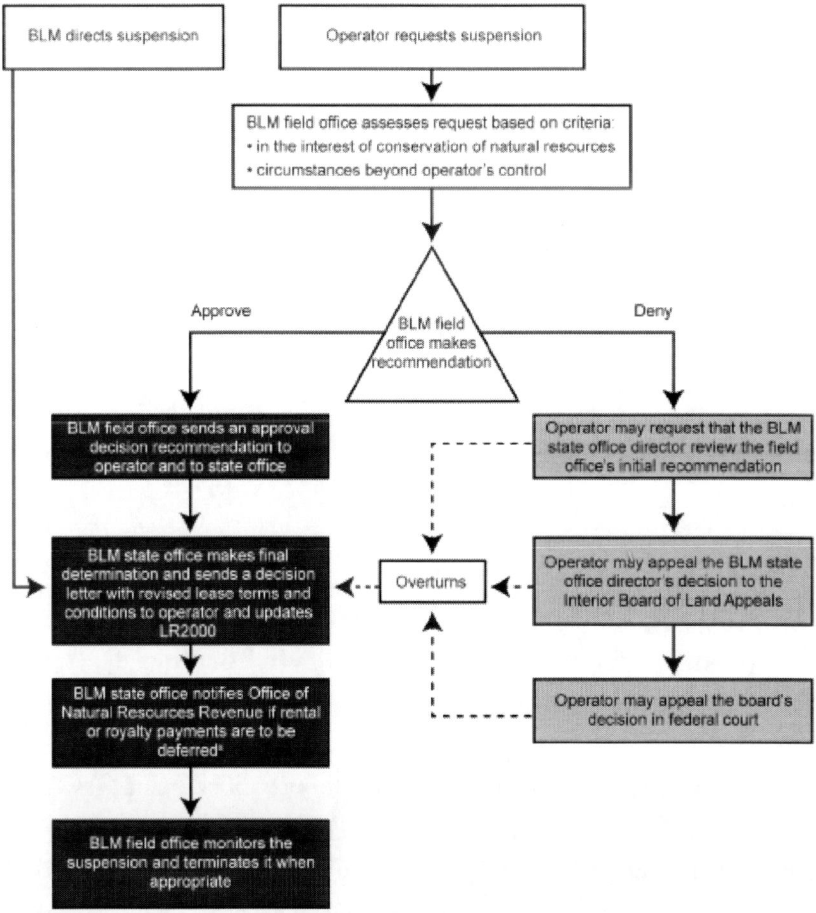

Source: GAO analysis of BLM Documentation. | GAO-18-411.

[a] The Office of Natural Resources Revenue, within the Department of the Interior, is responsible for collecting rental and royalty payments associated with leases on onshore federal lands. During a suspension of operations and production, both rental and minimum royalty payments are halted. Other types of suspensions do not halt collection of rental payments.

Figure 2. Typical Bureau of Land Management (BLM) Review Process for Oil and Gas Lease Suspension Requests.

A Small Portion of BLM's Oil and Gas Leases Were Recorded as Suspended, but Reasons for Suspensions Were Difficult to Determine

A small portion of BLM's oil and gas leases were suspended as of the end of fiscal year 2016, according to the agency's LR2000 data, but the reasons for the suspensions were difficult to determine. These data indicated that as of September 2016, about 2,750 of BLM's approximately 41,000 oil and gas leases were suspended in various locations for various lengths of time. LR2000 did not always contain the reasons for suspensions, which required us to take additional steps to identify the reasons.

As of September 2016, about 2,750 Oil and Gas Leases Were Recorded as Suspended and Varied in Their Location and Length of Suspension

According to LR2000 data, approximately 2,750 oil and gas leases were suspended at the end of fiscal year 2016. Our analysis of these data showed that the lease suspensions spanned 16 states and accounted for about 3.4 million acres of federally managed land. The data also showed that most of the suspensions were in five Mountain West states: Colorado, Montana, New Mexico, Utah, and Wyoming. These five states accounted for more than 2,350 of the approximately 2,750 recorded lease suspensions and encompassed more than 2.9 million acres of federally managed land (see app. II for additional details).[36] Figure 3, below, provides information on the location of oil and gas leases and recorded suspensions across the United States.

[36] According to BLM data, these five states also accounted for nearly 32,000 of BLM's 41,000 leases as of September 30, 2016.

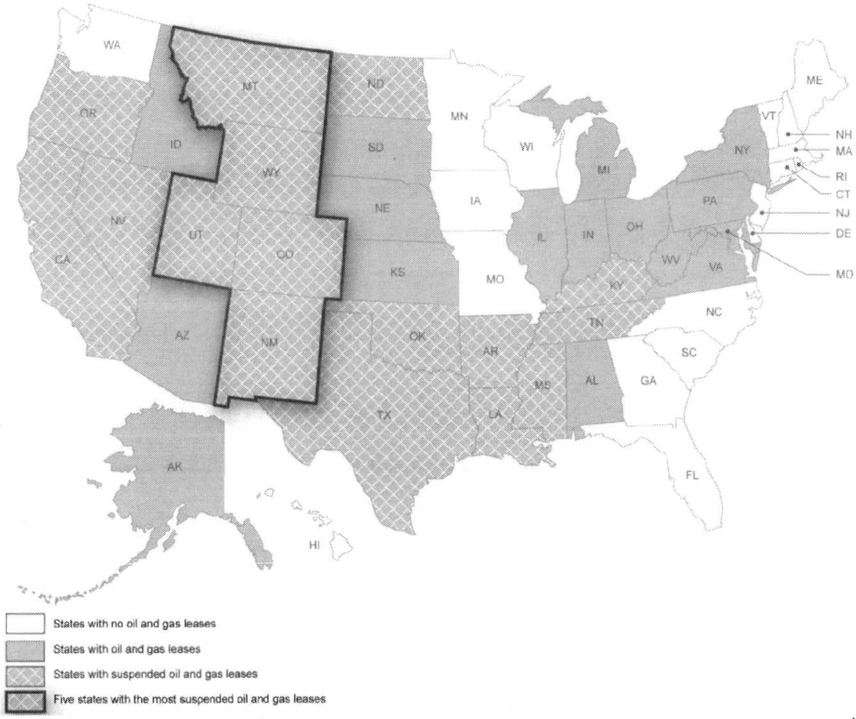

Source: GAO analysis of Bureau of Land Management data, Map resources (map). | GAO-18-411.

Figure 3. Location of Oil and Gas Leases and Recorded Suspensions, as of September 30, 2016.

Our analysis of LR2000 data showed that, of the approximately 2,750 recorded lease suspensions,

- about 630 had been in place for less than 3 years,
- about 1,150 had been in place for 3 years to less than 10 years,
- about 190 had been in place for 10 years to less than 20 years,
- about 130 had been in place for 20 years to less than 30 years, and
- about 650 had been in place for 30 years or more.

See figure 4 and appendix III for additional details.

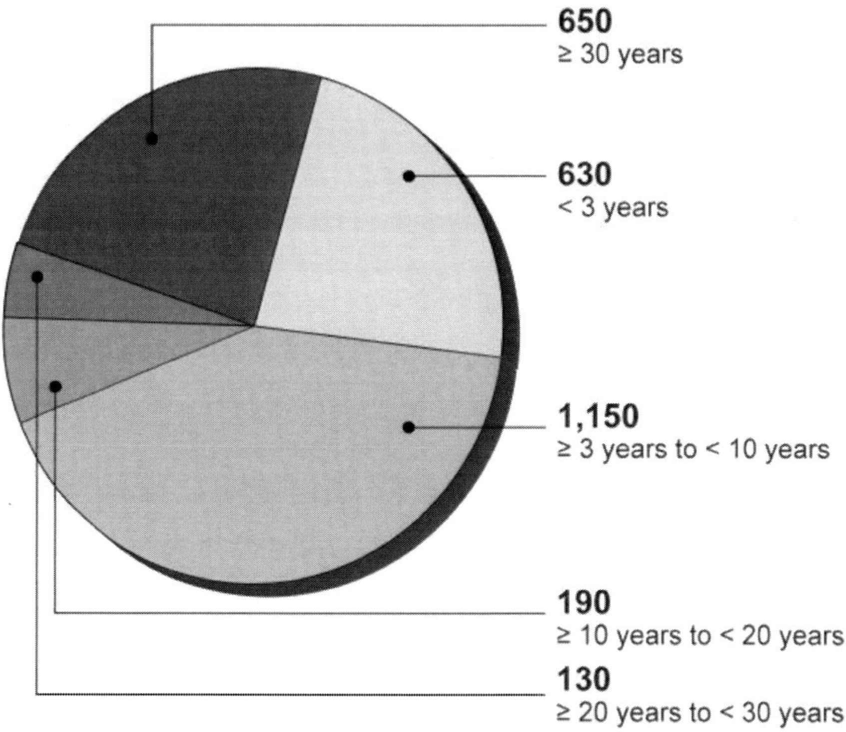

Source: GAO analysis of Bureau of Land management data. | GAO-18-411.

Note: Numbers represented in the figure have been rounded. The length of a lease suspension was calculated based on when the suspension was first issued through the end of fiscal year 2016— September 30, 2016.

Figure 4. Bureau of Land Management Oil and Gas Leases Recorded as in Suspension, by Length of Suspension, as of September 30, 2016.

Reasons for Suspensions Were Not Always Recorded in BLM's Database and Required Reviews of Official Lease Files to Identify

BLM's database, LR2000, did not always contain information on the reasons for oil and gas lease suspensions. BLM officials said that while LR2000 does not have a field to specifically capture the reason for a suspension, and inclusion of this information is not mandatory, the general remarks field could be used for this purpose.

Because we found this remarks field was rarely used to capture the reason for suspensions, we reviewed the official lease files for a sample of 48 leases in Montana and Wyoming that were suspended as of September 30, 2016, and we interviewed field office staff for clarification. The reasons for suspensions in this sample generally fell into four broad categories: environmental reviews, delays in reviewing applications for permits to drill, logistical conflicts, and other reasons.

Our review of the official lease files for our sample found the following reasons cited for suspensions:

- Sixteen leases were suspended for large-scale environmental concerns, such as wilderness or wildlife protection areas or environmental reviews that affected large parcels of land. These 16 suspensions had been in effect for approximately 6 years to 38 years. One of these leases was suspended because of a court order that also resulted in suspension of 422 other leases; the leases suspended as a result of this court order accounted for most of the suspensions that had been in place for more than 30 years.[37]
- Fourteen leases were suspended because BLM required additional time to complete its review of the operator's drilling permit application. These 14 suspensions had been in effect for approximately 1 year to 13 years. Seven of these 14 suspensions were issued because BLM needed additional time to review the environmental assessments submitted with the drilling permit applications.

[37] These leases were located on federal lands in Montana. The court ruled that no further action on the leases could occur without additional environmental and wildlife analyses. File documents from the Forest Service indicate that there was little interest at that time in conducting these analyses because of, among other things, their expense, and that there was also little interest from operators in developing these lands for oil and gas production. *Conner v. Burford,* 605 F. Supp. 107 (D. Mont. 1985), *aff'd in part and rev'd in part*, 836 F.2d 1521 (9th Cir. 1988), *superseded*, 848 F.2d 1441 (9th Cir. 1988), *cert denied sub. nom.,* Sun Exploration & Prod. Co. v. Lujan, 489 U.S. 1012 (1989).

- Eight leases were suspended because they faced logistical conflicts with other surface development, such as mining activities occurring on the lease or adjacent lands. These suspensions had been in effect for approximately 4 years to 25 years.
- Five leases were suspended for other, short-term reasons, such as weather-related issues or economic conditions, but were recorded in LR2000 as suspended for approximately 22 years to 74 years.[38]
- We were unable to determine the reasons why the 5 remaining leases were suspended. These leases were recorded as suspended for approximately 28 to 82 years. The agency was unable to provide lease files for 1 of the leases. Field officials said that some of these suspensions may have been issued at the state level, and the officials had no additional information on them.

According to *Standards for Internal Control in the Federal Government*, management should use quality information to achieve the entity's objectives; quality information may be defined as appropriate, current, complete, accurate, accessible, and provided on a timely basis.[39] BLM does not have quality information on the reasons for suspensions, in part because such reasons are not routinely included in LR2000, and there is no specific data field for them. To obtain this information, BLM officials would have to review the official lease files, as we did, and most of the files were available only in hard copy in BLM state offices. Therefore, the information is not readily accessible across the agency. Field officials we interviewed from one field office said that additional information on reasons for suspension in the database would be helpful in monitoring lease suspensions and in communicating with others, such as management or the public, about suspensions. BLM headquarters officials said they are planning to update or replace LR2000. By including a data field in the update or replacement for LR2000 to record the reasons for suspensions,

---

[38] As discussed later in this chapter, officials from relevant field offices indicated that these suspensions were not in effect for this period of time and that the information in LR2000 was outdated.

[39] GAO-14-704G.

BLM could better ensure that federal lands are not being inappropriately kept from development—potentially foregoing revenue—or from valuable uses of public lands.

BLM Relies on an Informal Monitoring Approach That May Not Provide for Consistent and Effective Oversight of Lease Suspensions

BLM uses an informal approach to monitor lease suspensions and does not have procedures in place for monitoring suspensions, which may not ensure consistent and effective oversight. We also found that BLM's state offices do not always maintain current information on lease suspensions in the official lease files or LR2000, and BLM headquarters and state officials told us they generally do not oversee the monitoring of lease suspensions.

Monitoring Varies Among Field Offices, and BLM Does Not Have Procedures for How to Conduct It

Field offices vary in how they monitor lease suspensions, and BLM does not have official agency procedures in place for monitoring, relying instead on an informal approach. We found that the field offices we reviewed differed in the frequency of their monitoring activities for lease suspensions.[40] According to officials we interviewed from these offices:

- 8 field offices monitor with varying frequency, depending on the circumstances; and
- 3 field offices monitor rarely.

[40] We reviewed a total of 12 field offices; however, 1 field office did not respond to our request for information on monitoring of lease suspensions.

Officials who monitored with varying frequency said that the frequency depends in part on the nature of the suspension. For instance, they said suspensions that involve seasonal protection of wildlife habitat, which can last for several months, typically require relatively little monitoring because the time frames for these suspensions are more clearly defined. In contrast, suspensions involving environmental reviews often require more frequent monitoring because the time frames associated with these suspensions are less definitive and can range from several months to several years. Several of these officials said that their offices have established prompts to alert staff when to conduct monitoring activities. For example, an official from 1 field office told us the office's staff use handwritten notes to track their lease suspensions. An official from another field office informed us that their office uses an electronic calendar feature to alert staff when to monitor, and several other field office officials reported that they rely on various spreadsheets and emails to remind them when to monitor. Officials from 1 field office also stated that their office uses an estimated end date for every suspension[41]—that is, the date the suspension is expected to terminate—to prompt them to review the current conditions to ensure that the suspension is still warranted.

Officials who monitored with varying frequency also said that the frequency depends on the availability of staff for monitoring. These officials said they generally rely on petroleum engineers in their respective offices to monitor lease suspensions because these individuals are normally the most familiar with leases.[42] However, some officials added that staffing limitations, particularly a shortage in petroleum engineers, have hindered their ability to monitor lease suspensions in a timely manner. Several of the field officials we interviewed noted that, in recent years, they have had to rely on other staff or petroleum engineers who were on loan from other field offices because their offices did not have a petroleum engineer on staff.

---

[41] End dates are estimated because it is difficult to predict the period of delay, and BLM's guidance states that suspensions should be granted for an indefinite term.

[42] Several field officials we interviewed informed us that they also rely on legal staff to assist with follow-up on lease suspensions, typically when legal documents are prepared for a lease, such as when the term and conditions of a lease are revised because of a suspension or in cases where a suspension has been terminated. They stated that these activities are usually one-time events and do not involve the routine monitoring of lease suspensions.

According to two field officials we interviewed, while assistance from other field offices is needed and appreciated, there is invariably a lack of consistency in the knowledge that engineers from other offices have about the lease sites involved. Field officials also said that there have been instances in which petroleum engineers left the agency for the private sector, resulting in a loss of institutional knowledge about certain leases, possibly contributing to lapses in follow-up on particular leases. Officials from offices that rarely or never conduct monitoring also cited problems with staff availability. We reported on human capital challenges at BLM, specifically in hiring and retaining petroleum engineers, in March 2010.[43] We also noted BLM's human capital constraints in our High-Risk Series update report in February 2011, and we reported on human capital issues at BLM in January 2014 and September 2016.[44] In several of these reports, we recommended that BLM take a number of actions, including using existing authorities and incentives to improve staff retention. BLM generally agreed with these recommendations and has taken action on some, but not all, of these recommendations.

Nonetheless, the extent of variability we found, including 3 field offices that monitor rarely or not at all, indicates that allowing individual field offices to determine when to monitor suspensions may not ensure that monitoring takes place. Under *Standards for Internal Control in the Federal Government*, management should design control activities, such as procedures, to ensure the objectives of the program are achieved.[45] The Office of Management and Budget has also acknowledged the importance of internal guidance documents to channel the discretion of employees, increase efficiency, and enhance the fair treatment of similarly situated

---

[43] GAO, *Oil and Gas Management: Interior's Oil and Gas Production Verification Efforts Do Not Provide Reasonable Assurance of Accurate Measurement of Production Volumes*, GAO-10-313 (Washington, D.C.: Mar. 15, 2010).

[44] GAO, *High-Risk: An Update*, GAO-11-278 (Washington, D.C.: February 2011); *Oil and Gas: Interior Has Begun to Address Hiring and Retention Challenges but Needs to Do More*, GAO-14-205 (Washington, D.C.: Jan. 31, 2014); and *Oil and Gas Oversight: Interior Has Taken Steps to Address Staff Hiring, Retention, and Training but Needs a More Evaluative and Collaborative Approach,* GAO-16-742 (Washington, D.C.: Sept. 19, 2016).

[45] GAO-14-704G.

parties.[46] Some field officials we interviewed said that procedures to help guide them on monitoring could be beneficial and provide a level of consistency. By developing procedures for monitoring lease suspensions, including when to conduct monitoring efforts, BLM could promote more consistent monitoring to better ensure that lease suspensions in effect are warranted.

BLM State Offices and Headquarters Generally Do Not Oversee Field Office Monitoring of Lease Suspensions and Do Not Always Have Current or Complete Information on Suspensions

Officials from BLM's state offices told us that they do not oversee field office monitoring of suspensions, and we found that they did not always have current or complete information on suspensions. We found that more than three-quarters of the official lease files in BLM state offices we reviewed contained outdated documentation regarding the status of lease suspensions. Specifically, files for 37 of the 48 lease suspensions we reviewed did not contain updated information on whether the lease suspension had been monitored or reviewed since the suspension was initially issued. For example, we reviewed a lease file for a suspension issued in 1949 for economic reasons, but the file only contained information on the suspension issuance and not whether monitoring occurred to assess the economic conditions associated with the lease. Additionally, we discovered that some official lease files were not complete and did not have certain required information, such as letters issuing the suspension. For example, three of the lease files we reviewed were missing required information. We could not verify the reasons these leases were suspended, their current status, or any information concerning monitoring efforts associated with them. Field officials we interviewed did not have any information on these suspensions and said that they may have been initiated by the state office more than 30 years ago. However, BLM

---

[46] Office of Management and Budget, *Final Bulletin for Agency Good Guidance Practices*, 72 Fed. Reg. 3432 (Jan. 25, 2007).

state officials were unable to confirm or deny this. For another lease, there was no lease file. Officials in BLM headquarters and state offices said that there is no requirement for them to oversee the field offices' monitoring activities. However, they said that performing such oversight could help to ensure effective and consistent monitoring of lease suspensions.

We also identified some instances in LR2000 where data on suspensions were not up to date. Specifically, 7 of the 48 leases we reviewed were recorded in LR2000 as suspended, but information we received from agency officials indicated that the suspensions were no longer warranted. We later confirmed with state and field officials that none of the 7 suspensions remained in effect. Five of these 7 leases were recorded as being in suspension for 22 years or more for what appeared to be short-term reasons, such as weather-related issues or economic conditions. One Wyoming suspension, for instance, was granted in 1990 because of low oil prices at the time, which made repairing wells uneconomical. While this lease was still recorded as suspended in LR2000 as of September 2016, a termination letter in the lease file indicated that the suspension was terminated in 1991. In another example, a lease was listed in the official lease file as suspended for 3 years because of delays in processing a drilling permit application. When we followed up with field officials about the lease, they informed us that the suspension should have been terminated years ago; however, we found no termination letter in the official lease file maintained by the state office. Field officials speculated that the letter may not have been sent because the case manager had retired and no one in the field office knew to follow up on the lease. Because field officials informed us that these leases were no longer suspended, we confirmed with officials from Interior's Office of Natural Resources Revenue that payments were being appropriately collected for these 7 leases.[47]

[47] The Office of Natural Resources Revenue tracks rents and royalty payments collected from oil and gas leases using a separate database. Office of Natural Resources Revenue officials informed us that their database does not electronically communicate with LR2000, and staff must manually convert identification numbers from LR2000 in order to cross check these leases against their database.

Moreover, these data are not available in a standardized report that could be used to help oversee monitoring, such as a report showing the average length or frequency of suspensions. BLM produces standardized reports from LR2000 for other aspects of oil and gas leases, such as when leases have been issued or are set to expire. BLM officials said that a standardized report for lease suspensions could assist headquarters and state officials in conducting oversight of field offices' monitoring efforts.

Standards for Internal Control in the Federal Government state that management should design control activities, such as conducting top-level reviews of actual performance, to ensure the objectives of the program are being achieved.[48] By requiring that management, particularly cognizant headquarters and state office officials, conduct top-level reviews of field offices' monitoring efforts, as well as top-level reviews of official lease files and databases, BLM could better ensure that lease suspensions in effect continue to be warranted and that information on suspensions is current and complete. Additionally, federal standards for internal control state that management should design control activities, such as developing mechanisms that enforce management's directives, to achieve the entity's objectives and address related risks. By developing mechanisms, such as summary reports on lease suspensions, as BLM updates or replaces LR2000, BLM could assist cognizant officials in headquarters and state offices with their oversight of monitoring.

CONCLUSION

The ability of federal agencies to manage their programs effectively depends in part on the information systems the agencies use and the quality of the data within these systems. Over the past several years, BLM has worked to improve the quality of the data in LR2000, including data related to oil and gas lease suspensions. These efforts have helped to improve the accuracy of certain data, but they do not address some

---

[48] GAO-14-704G.

constraints of LR2000. In particular, LR2000 does not contain a data field for recording the reasons for suspensions. BLM officials told us that they will upgrade or replace LR2000 in the near future. By including a data field in the update or replacement for LR2000 to record the reasons for suspensions, BLM could better ensure that federal lands are not being inappropriately kept from development—potentially foregoing revenue—or from other valuable uses of public lands.

BLM's ability to effectively manage the program also depends on the establishment of effective internal controls. To date, BLM has not developed procedures for monitoring lease suspensions. By developing procedures for monitoring lease suspensions, including when to conduct monitoring efforts, BLM could promote more consistent monitoring to better ensure that lease suspensions in effect are warranted. Additionally, BLM does not conduct top-level reviews to oversee field offices' monitoring efforts, and we found instances in which BLM's information on suspensions was outdated or incomplete. By requiring that management, particularly cognizant headquarters and state office officials, conduct reviews of field offices' monitoring efforts, as well as official lease files and databases, BLM could better ensure that information on suspensions is current and complete. Finally, BLM does not have mechanisms to provide officials with some key information relevant for oversight, such as when suspensions were last reviewed or the average length and frequency of suspensions. By developing mechanisms, such as summary reports on lease suspensions, as BLM updates or replaces LR2000, BLM could assist cognizant officials in headquarters and state offices with their oversight of monitoring.

Recommendations for Executive Action

We are making the following four recommendations to BLM:

As BLM updates or replaces its database, the Director of BLM should include a data field to record the reasons for suspensions. (Recommendation 1)

The Director of BLM should develop official agency procedures for monitoring oil and gas lease suspensions, including when to conduct monitoring activities. (Recommendation 2)

The Director of BLM should require cognizant officials in headquarters and state offices to conduct top-level reviews of field offices' monitoring of oil and gas lease suspensions, as well as of official lease files and databases to ensure they are current and complete. (Recommendation 3)

As BLM updates or replaces LR2000, the Director of BLM should ensure the development of mechanisms, such as standardized summary reports on lease suspensions, to assist cognizant officials in headquarters and state offices with oversight of field offices' monitoring efforts. (Recommendation 4)

AGENCY COMMENTS

We provided a draft of this chapter to Interior for review and comment. In its comments, reproduced in appendix IV, Interior generally agreed with our findings and recommendations. Interior also outlined plans for addressing the recommendations.

Regarding our first recommendation, Interior stated that it agrees that any future database used to track information on oil and gas lease suspensions should include a data field to more explicitly record the reasons for suspensions.

Interior also stated that it will develop standardized procedures for monitoring oil and gas lease suspensions, consistent with our second recommendation. These procedures will be instituted agency-wide, according to Interior, and agency policy and handbooks will be updated as needed to implement the procedures.

With respect to our third recommendation, Interior stated that it will provide updated guidance and online training to assist the state and field offices in managing, monitoring, and reviewing lease suspensions. These actions are positive steps and may address our recommendation depending on their implementation.

Finally, consistent with our fourth recommendation, Interior stated that any future update to or replacement of LR2000 database will include the capability to create standardized reports for oil and gas lease suspensions.

As agreed with your offices, unless you publicly announce the contents of this chapter earlier, we plan no further distribution until 30 days from the report date. At that time, we will send copies to the appropriate congressional committees, the Secretary of the Interior, and other interested parties.

Frank Rusco
Director,
Natural Resources and Environment

APPENDIX I: OBJECTIVES, SCOPE, AND METHODOLOGY

This chapter examines (1) the process the Bureau of Land Management (BLM) uses to determine whether to suspend oil and gas leases; (2) the extent of oil and gas lease suspensions and the reasons for the suspensions of selected leases; and (3) the approach that BLM uses to monitor the status of lease suspensions and the extent to which this approach allows for oversight of such lease suspensions.

To examine the process BLM uses to determine whether to suspend oil and gas leases, we reviewed applicable laws, agency documents, and the criteria BLM uses when considering lease suspensions. Specifically, we reviewed BLM's statutory requirements for granting a lease suspension. We also reviewed BLM's guidance for reviewing suspension requests—*Suspensions of Operations and/or Production Manual*—which outlines the process and criteria BLM uses to approve or deny a lease suspension request as well as the process for appealing a suspension decision. We interviewed BLM officials at headquarters, as well as state and field offices responsible for leases in our review, about how they apply these criteria when assessing suspension requests. We also interviewed representatives from the Interior Board of Land Appeals about suspension decisions that are appealed to the board, how these appeals are handled, board decisions that are subsequently appealed, and the process involved with those appeals.

To examine the extent of oil and gas lease suspensions and the reasons for the suspensions of selected leases, we analyzed data on lease suspensions from BLM's Legacy Rehost 2000 System (LR2000) database as of September 30, 2016.[49] We took a number of steps to assess the reliability of suspension data and related fields in LR2000. Specifically, we performed electronic tests to check the extent to which data were complete and within expected ranges. Testing included comparison of data extractions prepared by BLM officials for us against data we downloaded directly from LR2000. We also interviewed BLM officials responsible for managing the system about how data are collected and entered into the system as well as the steps the officials take to help ensure that the data are accurate and complete. We also clarified discrepancies regarding lease suspension data with these officials when necessary. We determined the data were sufficiently reliable to give a high-level summary on suspensions, including information on the number and location of leases, the number in suspension, and suspension length.

LR2000 contains information on activities related to an oil and gas lease's status, among other things. For each lease, we identified the latest record, if any, for actions in fiscal year 2016 and earlier that indicate suspension initiation or termination. We determined that a lease was in suspension if the most recent action related to a suspension indicated that the suspension was initiated. We determined the length of suspension based on the date of that initiation record. We then determined distributions of the numbers of leases recorded as still in suspension in each state as of the end of fiscal year 2016.

We also reviewed the official lease files, maintained by BLM state offices, for a nongeneralizable sample of leases recorded as suspended as of September 30, 2016, in Montana and Wyoming to assess their status and

---

[49] LR2000 contains information on all mineral and land assets managed by BLM, including oil and gas leases.

the reasons behind the suspensions. We chose these two states because they were among the states with the largest numbers of suspended leases. Montana's official lease files were electronically maintained and easily accessible, while Wyoming, which had leases recorded as suspended for the longest period of time as of September 30, 2016, maintained hard copy official lease files. Montana and Wyoming collectively represent about 50 percent of all oil and gas leases recorded as suspended. We used the following approaches to select a sample of 48 suspended leases in these states and limited the extent to which we selected multiple leases that were suspended at the same time for the same reason.

For Montana leases, we found that only 12 suspension initiation dates were recorded for the leases in suspension as of the end of fiscal year 2016. We therefore randomly selected for review a single lease from those suspended on each of these dates. For Wyoming, the suspension initiation dates were much more dispersed, so we identified groups of 15 or more leases based on a combination of suspension date, similarity of lease numbers, and the field office of jurisdiction. From these groups, we selected 19 suspended leases—each lease was the lease with largest acreage from each field office within its group. There were a number of leases that did not fit into these groups because there were fewer than 15 suspensions on a given date with similar lease numbers, so we selected a single lease file with the largest acreage from each year that was at least 20 years old. This allowed us to review suspensions that have been in effect for a relatively long period of time. This approach resulted in our selection of an additional 17 suspended leases in Wyoming. While our review of suspended lease files is not generalizable to other BLM lease suspensions, our findings provide examples of types of reasons that are cited for lease suspensions.

To verify the status of each selected lease, we compared information in LR2000 and the official lease file to information in the Offices of Natural Resources Revenue's database. The Office of Natural Resources Revenue, within the Department of the Interior, is responsible for collecting rental and royalty payments associated with oil and gas leases. We also interviewed the BLM state and field office officials responsible for the specific lease files we reviewed to obtain additional information about the status of certain lease suspensions and the reasons these suspensions remained in effect. We compared how BLM maintains and verifies its lease suspension information with *Standards for Internal Control in the Federal Government* for information and communication.[50]

To examine the approach BLM uses to monitor the status of lease suspensions and the extent to which the approach provides for oversight, we reviewed agency data, guidance and requirements, and official lease documents. In particular, we reviewed monitoring information in LR2000, BLM's *Suspensions of Operations and/or Production Manual*, and monitoring information in the official lease files for our sample of 48 leases recorded as being in suspension as of September 30, 2016. We also interviewed officials from BLM headquarters, as well as BLM's state offices in Montana and Wyoming and the field offices responsible for the 48 selected leases in our review—a total of 12 field offices, 2 from Montana and 10 from Wyoming. We interviewed officials from 11 field offices about the approaches they used to monitor lease suspensions, including the frequency of monitoring and the staff involved.[51] We also interviewed officials at headquarters and state offices to examine the extent to which these approaches provided for oversight of lease suspensions. We compared BLM's actions and documentation with agency guidance, federal regulations, and federal standards for internal control for control activities.[52]

We conducted this performance audit from September 2016 to June 2018 in accordance with generally accepted government auditing standards. Those standards require that we plan and perform the audit to obtain sufficient, appropriate evidence to provide a reasonable basis for our findings and conclusions based on our audit objectives. We believe that the evidence obtained provides a reasonable basis for our findings and conclusions based on our audit objectives.

---

[50] GAO, *Standards for Internal Control in the Federal Government*, GAO-14-704G (Washington, D.C.: September 2014).

[51] We reviewed a total of 12 field offices; however, 1 office did not respond to our request for information on monitoring of lease suspensions.

[52] GAO-14-704G.

We conducted this performance audit from September 2016 to June 2018 in accordance with generally accepted government auditing standards. Those standards require that we plan and perform the audit to obtain sufficient, appropriate evidence to provide a reasonable basis for our findings and conclusions based on our audit objectives. We believe that the evidence obtained provides a reasonable basis for our findings and conclusions based on our audit objectives.

APPENDIX II: BUREAU OF LAND MANAGEMENT OIL AND GAS LEASES, SUSPENSIONS, AND ACREAGE, AS OF SEPTEMBER 30, 2016

Table 1. Bureau of Land Management (BLM) Oil and Gas Recorded Leases, Suspensions, and Acreage, as of September 30, 2016

State	Number of leases	Leased acreage	Number of suspensions	Suspended acreage
Alabama	114	70,620	0	0
Alaska	174	1,006,960	0	0
Arizona	6	12,217	0	0
Arkansas	619	416,194	138	147,990
California	530	198,820	27	20,870
Colorado	4,135	3,099,068	180	236,901
Idaho	15	19,852	0	0
Illinois	8	1,581	0	0
Indiana	8	11,358	0	0
Kansas	462	120,151	0	0
Kentucky	69	38,073	4	7,516
Louisiana	578	349,457	2	133
Maryland	4	2,637	0	0
Michigan	235	129,769	0	0
Mississippi	630	351,925	2	1,025
Montana	2,751	2,145,774	607	992,145
Nebraska	27	10,442	0	0
Nevada	731	1,385,511	10	13,663
New Mexico	8,021	4,508,477	158	84,256
New York	5	1,183	0	0

Table 1. (Continued)

State	Number of leases	Leased acreage	Number of suspensions	Suspended acreage
North Dakota	1,854	889,427	88	68,345
Ohio	259	55,847	0	0
Oklahoma	1,209	250,358	2	440
Oregon	111	187,197	105	172,759
Pennsylvania	75	9,954	0	0
South Dakota	383	168,938	0	0
Tennessee	2	736	1	620
Texas	676	406,733	7	12,288
Utah	3,045	2,984,294	731	1,054,288
Virginia	38	27,509	0	0
West Virginia	242	100,917	0	0
Wyoming	13,996	9,195,483	687	568,775
Total	41,003	28,157,462	2,749	3,382,014

Source: GAO analysis of BLM data. | GAO-18-411.

Note: We reviewed BLM data to identify leases and suspensions and associated acreage for all states with BLM leases.

APPENDIX III: BUREAU OF LAND MANAGEMENT OIL AND GAS LEASES RECORDED AS IN SUSPENSION

Table 2. Bureau of Land Management (BLM) Oil and Gas Leases Recorded as in Suspension by Length and State, as of September 30, 2016

State	< 3 years	≥ 3 years to < 10 years	≥ 10 years to < 20 years	≥ 20 years to < 30 years	≥ 30 years	Total
Arkansas	11	125	0	2	0	138
California	22	2	1	2	0	27
Colorado	59	113	5	0	3	180
Kentucky	0	0	0	4	0	4
Louisiana	0	1	1	0	0	2
Mississippi	1	0	0	1	0	2
Montana	12	167	1	3	424	607
North Dakota	66	22	0	0	0	88
New Mexico	49	15	4	11	79	158
Nevada	2	1	3	4	0	10

State	< 3 years	≥ 3 years to < 10 years	≥ 10 years to < 20 years	≥ 20 years to < 30 years	≥ 30 years	Total
Oklahoma	0	2	0	0	0	2
Oregon	105	0	0	0	0	105
Tennessee	0	0	1	0	0	1
Texas	0	6	1	0	0	7
Utah	160	326	111	29	105	731
Wyoming	142	374	67	70	34	687
Total	629	1,154	195	126	645	2,749

Source: GAO analysis of BLM data. | GAO-18-411.

Note: We reviewed BLM data to identify the length of suspensions for all states with suspended leases. We calculated the length of a lease suspension based on when the suspension was first issued through the end of fiscal year 2016—September 30, 2016.

Appendix IV: Comments from the Department of the Interior

United States Department of the Interior

OFFICE OF THE SECRETARY
Washington, DC 20240

MAY 22 2018

Mr. Frank Rusco
Director
Natural Resources and Environment
U.S. Government Accountability Office
441 G Street NW
Washington, DC 20548

Dear Mr. Rusco:

Thank you for giving the Department of the Interior the opportunity to review and comment on the Government Accountability Office's (GAO) draft report entitled, *Oil and Gas Lease Management: BLM Could Improve Oversight of Lease Suspensions with Better Data and Monitoring Procedures* (GAO-18-411).

The draft report includes four recommendations for the Bureau of Land Management (BLM). The Department generally agrees with the overall findings of the draft report and concurs with all four recommendations included in the report. A summary of BLM's responses and plans for addressing the recommendations are provided below.

Recommendation 1: As BLM updates or replaces its database, the Director of BLM should include a data field to record the reasons for suspensions.

Response: The BLM concurs with this recommendation. While the current version of the BLM's Legacy Rehost 2000 (LR2000) database includes an action remarks field that used to record the reasons for lease suspensions, BLM agrees that any electronic database to track information on oil and gas lease suspensions should include a data field to more explicitly record the reasons for suspensions.

Recommendation 2: The Director of BLM should develop official agency procedures for monitoring oil and gas lease suspension, including when to conduct monitoring activities.

Response: The BLM concurs with this recommendation. The BLM will develop improved and standardized procedures for monitoring oil and gas lease suspensions, which will be applied consistently across the Bureau. The BLM will implement these improved lease suspension monitoring practices through written policy and update handbooks as necessary to provide practical implementation guidance.

Recommendation 3: The Director of BLM should require cognizant officials in headquarters and state offices to conduct top-level reviews of field offices' monitoring of oil and gas lease suspension, as well as official lease files and databases to ensure they are current and complete.

Response: The BLM concurs with this recommendation. The BLM's Washington Office and National Operations Center currently use a "DataFlux" tool, which compares LR2000 data with the existing data standard and business rules, to generate reports generally regarding each BLM state's LR2000 data accuracy, completeness, consistency, and validity, to include oil and gas lease suspensions.

In addition to using the DataFlux tool to review the quality of electronic data on lease suspensions, BLM will also update H-3103-1, the *Fees, Rentals, and Royalty* handbook and develop online training for lease suspensions to provide updated guidance and training resources to assist the state and field offices in managing, monitoring, and reviewing lease suspensions.

Recommendation 4: As BLM updates or replaces LR2000, the Director of BLM should ensure the development mechanisms, such as standardized summary reports on lease suspensions, to assist cognizant officials in headquarters and state offices with oversight of field offices' monitoring efforts.

Response: The BLM concurs with this recommendation. The BLM will update and improve its LR2000 reporting capabilities by developing a standardized report for oil and gas leases in a suspended status. When BLM updates or replaces LR2000, BLM will also ensure that any data system updates or replacements include the capability to create standardized reports for oil and gas lease suspension data.

Sincerely,

Joseph R. Balash
Assistant Secretary
Land and Minerals Management

In: Focus on Oil and Gas
Editor: Olivia B. Wilder
ISBN: 978-1-53616-325-4

Chapter 3

OIL AND GAS DEVELOPMENT: ACTIONS NEEDED TO IMPROVE OVERSIGHT OF THE INSPECTION AND ENFORCEMENT PROGRAM*

United States Government Accountability Office

ABBREVIATIONS

BIA	Bureau of Indian Affairs
BLM	Bureau of Land Management
ONRR	Office of Natural Resources Revenue
PET	petroleum engineer technician

* This is an edited, reformatted and augmented version of the United States Government Accountability Office Report to the Chairman, Committee on Natural Resources, House of Representatives, Publication No. GAO-19-7, February 2019.

Why GAO Did This Study

BLM has primary responsibility for managing oil and gas development on federal and Indian lands. To help ensure operator compliance with laws and regulations, BLM administers the Inspection and Enforcement program. Under the program, BLM inspects operators' drilling, production, and plugging activities and can issue various enforcement actions, such as monetary assessments, for violations. GAO was asked to examine key aspects of the Inspection and Enforcement program.

This chapter (1) describes the distribution of BLM's oil and gas Inspection and Enforcement program's workload and workforce among agency field offices for the most recent 5 years for which such data were available (fiscal years 2012 through 2016) and (2) examines the extent to which BLM conducted internal control reviews in accordance with its July 2012 oversight policy for fiscal years 2013 through 2018, the most recent period for which such data were available. GAO examined BLM policies, data, and documents; interviewed BLM headquarters, state and field office officials; visited six BLM field offices selected based on their level of resource development activity; and toured oil and gas drilling, production, and plugging sites at three of these six field offices.

What GAO Recommends

GAO is making three recommendations to BLM, including taking actions to increase monitoring of state offices' progress toward completing internal control reviews. BLM concurred with all three recommendations.

What GAO Found

Based on GAO's analysis of Bureau of Land Management (BLM) data, the distribution of BLM's oil and gas Inspection and Enforcement program's workload and workforce showed an imbalance among BLM's 33 field offices in fiscal years 2012 through 2016. GAO analyzed BLM data on the overall percentage of the workload and workforce distributed at each field office (i.e., activity level) and grouped similar activity level field offices together into highest, medium and lowest activity categories. GAO found that the program distributed the majority of its workload to 6 highest activity offices and distributed the majority of the workforce to 21 medium activity offices (see fig.). Based on GAO's review of BLM documentation and interviews with agency officials, BLM took both short- and long-term actions in fiscal years 2012 through 2016 to address this imbalance, such as temporarily re-assigning inspectors from some medium activity offices to some of the highest activity offices.

BLM has not completed all required internal control reviews of its field offices. BLM's July 2012 oversight policy instructs its state offices to periodically conduct internal control reviews of field offices, which are to, among other things, identify staffing needs. BLM state offices completed internal control reviews at 6 of 33 field offices from 2013 through 2017, and 5 more are scheduled from 2018 through 2020. Officials from BLM state offices told GAO that some human capital and workload challenges hindered their ability to complete reviews, including long-term vacancies in some state offices positions. However, a senior BLM official said that headquarters did not consistently track and monitor the extent to which state offices completed field office internal control reviews, and headquarters officials said they were not aware that so few reviews had been completed. Under federal standards for internal control, management should design control activities to achieve objectives and respond to risks, such as by comparing actual performance to expected results and analyzing significant differences. Identifying the reasons it did not complete internal control reviews, developing and implementing a plan to address those challenges, and monitoring state offices' progress toward completing required reviews will better position BLM to ensure that its state offices are completing all required internal control reviews as called for by its July 2012 oversight policy.

February 14, 2019

The Honorable Raúl Grijalva
Chairman
Committee on Natural Resources
House of Representatives

Dear Mr. Chairman:

Production of oil and gas on federal and Indian lands is an important part of the nation's energy portfolio and generates billions of dollars annually for the federal government, tribes, and individual Indian mineral owners. In fiscal year 2016, the Department of the Interior reported collecting more than $2 billion associated with onshore oil and gas development on federal and Indian lands, including royalty payments from operators for resources extracted. Interior's Bureau of Land Management (BLM) has primary responsibility for managing and overseeing oil and gas development on federal lands.[1] BLM also plays a key role in managing oil and gas development on Indian lands.[2] Specifically, BLM's Inspection and Enforcement program ensures production accountability (i.e., operators accurately measure and report production volume—these volumes are a key input into royalty payment calculations), environmental protection, and public safety.

Inspecting oil and gas development and production is a complicated process that requires BLM's petroleum engineer technician (PET) inspectors to conduct a variety of technically challenging inspections in

---

[1] These lands include about 700 million sub-surface mineral acres held by BLM, the U.S. Forest Service, and other federal agencies and surface owners.

[2] With respect to oil and gas leasing and development on Indian trust and restricted lands, BLM generally reviews, approves, and issues documents required for development, such as drilling permits and revenue-sharing agreements, in consultation with and with concurrence from Interior's Bureau of Indian Affairs (BIA), and conducts inspection and enforcement activities on such lands. Indian trust and restricted lands for these purposes mean those lands for which title is held in trust by the federal government for the beneficial interest of the tribe or a member and those lands for which title is held by the tribe or a member subject to restrictions on alienation. Such lands generally cannot be leased without approval of the Secretary of the Interior, who has generally delegated this authority to BIA.

geographically remote areas to interact with hundreds of different operators in the oil and gas market. At 33 BLM field offices with ongoing oil and gas development activities, the Inspection and Enforcement program employs a workforce of PET inspectors who carry out a workload that includes a range of inspections that cover a well's life cycle, from drilling and production to plugging. PET inspectors play a vital role in helping to ensure operators employ sound and prudent industrial practices at more than 140,000 oil and gas wells on federal and Indian lands. In addition, changes and innovations within the oil and gas industry can impact the bureau's PET inspector workforce and its inspection workload. In this context, BLM oversight is essential to help ensure the bureau's field offices have effective internal controls, which help BLM achieve its mission and respond to changing risks.

Past work by us, Interior's Office of Inspector General, and others has highlighted the importance of BLM's oversight of oil and gas development on federal and Indian lands and identified a number of weaknesses. BLM has taken some actions to address these weaknesses. For example:

- In March 2010, we found BLM was unable to consistently meet its goals for completing production inspections.[3] We recommended that BLM consider an alternative inspection strategy to ensure it inspects all wells within a reasonable time frame, given available resources. BLM concurred with our recommendation. In response to our recommendation, in 2011, BLM developed its risk-based production inspection strategy, which classifies wells as either high-priority or low-priority for inspections based on multiple factors.
- In December 2010, the Interior Inspector General identified weaknesses with the quality of inspections conducted by Inspection and Enforcement program personnel and made a number of recommendations. In July 2012, BLM issued a program oversight

[3] GAO, *Oil and Gas Management: Interior's Oil and Gas Production Verification Efforts Do Not Provide Reasonable Assurance of Accurate Measurement of Production Volumes*, GAO-10-313 (Washington, D.C.: Mar. 15, 2010).

policy that, among other things, requires state offices to conduct periodic internal control reviews of the Inspection and Enforcement program at BLM field offices with ongoing oil and gas activities.[4]

- Since 2011, we have designated Interior's management of federal oil and gas resources as a high-risk area vulnerable to fraud, waste, abuse, and mismanagement.[5] In our February 2017 high-risk series update, we reported that Interior has taken some actions to address production verification and human capital challenges by, for example, using special, higher salary rates to hire and retain key oil and gas staff such as BLM inspectors.[6] However, the agency's management of federal oil and gas resources remains an area vulnerable to fraud, waste, abuse, and mismanagement.
- In our 2017 high-risk series update, we added a new high-risk area on improving management of federal programs that serve tribes and their members. We reported that, among other things, Interior's Bureau of Indian Affairs mismanages Indian energy resources held in trust and thereby limits opportunities for tribes and their members to use those resources to create economic benefits and improve the well-being of their communities.

You asked us to examine BLM's oil and gas Inspection and Enforcement program. This chapter (1) describes the distribution of oil and gas Inspection and Enforcement program workload and workforce among BLM's field offices for the past 5 years and (2) examines the extent to which BLM conducted internal control reviews in accordance with its July 2012 oversight policy.

To describe the distribution of oil and gas inspection and enforcement workload and workforce among BLM's 33 field offices with ongoing oil and gas activities, we focused on the workload completed by PET

---

[4] U.S. Department of the Interior, Bureau of Land Management, Oversight of the Oil and Gas Inspection and Enforcement Program, Information Memorandum No. 2012-161 (Washington, D.C.: July 27, 2012).

[5] GAO, *High-Risk Series: An Update*, GAO-11-278 (Washington, D.C.: Feb. 16, 2011).

[6] GAO, *High-Risk Series: Progress on Many High-Risk Areas, While Substantial Efforts Needed on Others,* GAO-17-317 (Washington, D.C.: Feb. 15, 2017).

inspectors, as well as workforce data on PET inspectors. PET inspectors' workload includes inspections of drilling, production, and well plugging operations, commonly referred to as "downhole" inspections according to the BLM Inspection and Enforcement Program Manager.[7] We focused on these areas because (1) BLM planning documents consistently state that downhole inspections on high-priority drilling, high-priority well plugging, and high-priority production wells are the program's top three work activities, and (2) the program allocates the majority of its total inspection work months to completing downhole inspections.[8] We selected the past 5 years (fiscal years 2012 through 2016) for which such workload and workforce data were available during our review.

However, due to data reliability issues associated with priority rankings (high or low) during fiscal years 2012 through 2016 for drilling, production, and plugging inspections as well as issues with the number of completed inspections—the most direct measure of workload—we determined that these inspection data were not sufficiently reliable for our purposes.[9] Instead, we identified seven alternate workload measures for

---

[7] We excluded from our scope other types of inspections performed by PET inspectors that represent a small percentage of BLM's inspection workload, such as idle well inspections. We also excluded inspections conducted primarily by production accountability technicians—staff responsible for ensuring that production of oil and gas is accurately reported—such as production records verification inspections because these also represent a small percentage of the inspection workload. Finally, we excluded inspections conducted by environmental protection and natural resource specialists intended to verify operators' compliance with certain lease and permit requirements related to surface environment, due to recent GAO work on the subject. See *Oil and Gas Development: BLM Needs Better Data to Track Permit Processing Times and Prioritize Inspections*, GAO-13-572 (Washington, D.C.: Aug. 23, 2013) and *Oil and Gas Development: Improved Collection and Use of Data Could Enhance BLM's Ability to Assess and Mitigate Environmental Impacts*, GAO-17-307 (Washington, D.C.: Apr. 25, 2017).

[8] According to BLM officials, a work month is about 172 hours and is calculated by dividing 2,080 (i.e., the total number of hours generally worked by a full-time employee in a year) by 12 months.

[9] In December 2010, the Interior Inspector General reported that BLM cannot completely verify which wells have been inspected due to data integrity weaknesses with its corporate oil and gas database. U.S. Department of the Interior, Office of the Inspector General, *Bureau of Land Management's Oil and Gas Inspection and Enforcement Program*, CREV-BLM-0001-2009, (Washington, D.C.: December 2010). To address this issue, BLM planned to develop a revised inspection record system by September 30, 2016. According to BLM documentation, the revised inspection record system would provide data entry controls and automate some data entry, which would increase the reliability of inspection data. As of August 2018, BLM had not deployed the revised inspection record system, and agency

drilling, production, and plugging inspections, which BLM officials said would provide a representation of the size and scope of its inspection workload. In addition, we identified two workforce measures for a total of nine measures that covered fiscal years 2012 through 2016. The workload measures are:

- the number of wells drilled on federal and Indian lands;
- the number of production cases on federal and Indian lands (a case is either a lease or an agreement, can have from 1 to more than 1,000 wells, and is the unit of analysis for BLM's production inspection workload);
- the number of high-priority production cases on federal and Indian lands;
- the number of low-priority production cases on federal and Indian lands;
- the number of planned work months for plugging inspections on federal and Indian lands;[10]
- the number of enforcement actions issued for all inspections on federal and Indian lands;[11] and
- the number of major and continued noncompliance enforcement actions—major incidents of noncompliance, monetary assessments, civil penalties, and operation shutdown—issued for

officials said they had a deployment goal of deploying the system during the second quarter of fiscal year 2019. BLM officials said that data related to the number of high-priority drilling inspections completed during fiscal years 2012, 2013, and 2014 would not be reliable because BLM did not properly record the inspection priority level (high or low) for a large number of drilling inspections. BLM took action in fiscal year 2015 to conduct more frequent and detailed reviews of these data. Also, with respect to high-priority plugging inspections, we found that BLM's corporate oil and gas database did not have the ability to record and track a plugging inspection as either high or low priority. BLM officials said that this capability will be included in the revised inspection record system.

[10] We did not analyze the extent to which the actual number of work months BLM spent on plugging inspections for these years matched or differed from the number of planned plugging inspection work months recorded in BLM's planning documents.

[11] According to BLM officials, the bureau's corporate oil and gas database does not link enforcement actions to the specific inspection that identified the violation. It is not possible to analyze enforcement action data issued as a result of downhole inspections alone. As such, the enforcement actions data include all such actions issued during all types of inspections (i.e., downhole, environmental, and records verification inspections).

all inspections on federal and Indian lands. These types of actions generally identify (1) immediate, substantial, and adverse effects to public health and safety, the environment, production accountability, or royalty income; or (2) an operator's failure to correct an identified problem or violation by the abatement date. For example, BLM can issue monetary assessments up to $5,000 to operators who drill on federal and Indian lands without approval.

- With respect to workforce, the measures are:
- the number of PET inspectors "on board" at each field office at the start of each fiscal year,[12] and
- the number of PET inspection work months in each fiscal year.

For each of the workforce and workload measures, we calculated the percentage of the nationwide total (i.e., federal plus Indian) for each field office during fiscal years 2012 through 2016. Our calculations, in general, demonstrated a pattern that showed natural break points in the data for field office activity level. Based on these natural break points, we developed and applied decision rules. Based on these rules, we categorized each field office, based on specific workload and workforce measures, as "high activity" if it represented 4 percent or more of the nationwide total, "medium activity" if a field office represented from 3.9 percent to 1 percent of the nationwide total, and "low activity" if a field office

---

[12] We focused on the number of PET inspectors "on board" at the start of each fiscal year because BLM uses such data as the planning basis for the number and types of inspections to be completed during the fiscal year. The total number of authorized PET inspector positions at each field office is calculated by adding onboard personnel and vacant positions. We did not assess BLM efforts to fill vacant positions due to recent work on this subject. Specifically, we previously reported that Interior experiences difficulties with filling vacancies for key oil and gas personnel, including PET inspectors. See *Oil and Gas: Interior Has Begun to Address Hiring and Retention Challenges but Needs To Do More*, GAO-14-205, (Washington, D.C.: Jan. 31, 2014). We also previously reported that BLM had taken steps to resolve its hiring challenges, including the use of special salary rates, incentive payments, and student loan repayments. See *Oil and Gas Oversight: Interior Has Taken Steps to Address Staff Hiring, Retention and Training Needs but Needs a More Evaluative and Collaborative Approach,* GAO-16-742 (Washington, D.C.: Sep. 29, 2016).

represented 1 percent or less of the nationwide total.[13] BLM agreed with this approach for categorizing the 33 field offices based on their workload and workforce activity levels, and said it was representative of the workload and workforce. For example, we categorized the Rawlins, Wyoming, field office as medium activity for the number of wells drilled because the office was responsible for 1.8 percent of all wells drilled on federal and Indian lands from fiscal years 2012 through 2016. However, we categorized the Rawlins field office as high activity based on the number of planned work months for plugging inspections because the office represented 5.7 percent of planned work months for plugging inspections on federal and Indian lands in fiscal years 2012 through 2016.

We then compared the distribution of field offices across the workload and workforce measures to identify the offices that most consistently fell into the high-activity, medium-activity, or low-activity category. Once again, our analysis demonstrated a pattern that showed natural break points for the data, which allowed us to develop certain decision rules. Based on these rules, we categorized a field office as an overall highest activity office if it was rated high activity in at least six of the nine workload and workforce measures and an overall lowest activity office if it was rated low activity in at least six of the nine measures. Overall medium-activity offices were those that did not fall into the overall high and overall low categories. For example, we categorized the Rawlins field office as an overall medium-activity office because it was rated as high activity for three measures and medium activity for six measures. BLM agreed with this approach for grouping the field offices into three categories based on activity.

We obtained our workload and workforce data primarily from the annual inspection work plan BLM develops at the start of each fiscal year. The annual work plan provided data for production cases, planned work

[13] In some instances, it did not make sense to use an exact cutoff percentage because such a cutoff would skew the distribution. For example and for the "number of major and continued noncompliance enforcement actions" workload measure, an exact 1 percent cut off would have put 16 field offices (or almost half of all field offices) into the lowest activity category. Instead, we used 0.7 percent as the cut off, which put 11 field offices (or one-third of all field offices) into the lowest activity category.

months for plugging inspections, number of onboard PET inspectors, and number of PET inspection work months. In addition, we obtained records-level data from the bureau's corporate oil and gas database for the wells drilled and enforcement actions data. We assessed the reliability of the work plan and records-level data by reviewing related documentation, interviewing knowledgeable agency officials, and conducting electronic and manual testing of the data to identify missing data or obvious errors. We found that the data related to our seven workload and two workforce measures were reliable for the purposes of our review.

To determine the extent to which BLM conducted internal control reviews of the Inspection and Enforcement program from fiscal years 2013 through 2018 in accordance with its July 2012 oversight policy, we reviewed the oversight policy and compared its requirements to (1) the field office internal control reviews completed by state office officials and (2) federal standards for internal control, specifically the control activities standard that requires management to design control activities to achieve objectives and respond to risks.[14] We selected this time period as it was the most recent period for which data were available.

For both objectives and to obtain possible reasons or explanations for the data, we reviewed key BLM policy and guidance documents, including the *Inspection and Enforcement Handbook* and the annually issued *Inspection and Enforcement Strategy and Goals*. We interviewed BLM headquarters officials, BLM officials at 7 state offices, and PET inspectors at 13 BLM field offices. We selected this nongeneralizable sample to provide a range of resource development, such as the primary type of resource developed (i.e., oil or gas or both), the presence of Indian lands, and geographic variation. We also conducted site visits to six BLM field offices in New Mexico, Wyoming, and North Dakota; we selected these field offices as they have historical or current high levels of resource development activity. During those site visits, we accompanied PET inspectors on drilling, production, and plugging inspections.

---

[14] GAO, *Standards for Internal Controls in the Federal Government*, GAO-14-704G (Washington, D.C.: Sept. 10, 2014).

We conducted this performance audit from December 2016 to February 2019 in accordance with generally accepted government auditing standards. Those standards require that we plan and perform the audit to obtain sufficient, appropriate evidence to provide a reasonable basis for our findings and conclusions based on our audit objectives. We believe that the evidence obtained provides a reasonable basis for our findings and conclusions based on our audit objectives.

BACKGROUND

This section provides information on BLM's mission and organizational structure, the process for overseeing the development of federal and Indian oil and gas resources, and key aspects of the Inspection and Enforcement program.

BLM Mission and Organizational Structure

BLM's mission is to maintain the health, diversity, and productivity of public lands for present and future generations. As part of this mission, BLM manages federal lands for multiple uses, including recreation; grazing; timber; minerals; watershed; wildlife and fish; natural scenic, scientific, and historical preservation; and the sustained yield of renewable resources. BLM manages these responsibilities through its headquarters office in Washington, D.C.; state offices; district offices; and field offices. Each level's general responsibilities include the following:

- BLM's headquarters office develops guidance and regulations.
- State and field offices manage and implement the bureau's programs. In addition to implementing programs, BLM state offices oversee field office operations. Field offices lead BLM's oversight of oil and gas development. They are located primarily

in the Mountain West, where much of oil and gas development on federal and Indian lands takes place.

- Within field offices, BLM supervisory and staff PET inspectors and tribal PET inspectors (who are contracted by BLM to inspect some wells on Indian lands in accordance with BLM policies and procedures) have primary responsibility for implementing the Inspection and Enforcement Program with assistance from state office program coordinators, according to the Inspection and Enforcement Program Manager. Among other things, state office program coordinators help field offices plan and prioritize their inspection workloads in accordance with BLM policy and comply with BLM guidance and federal regulations when conducting and documenting inspections, according to BLM officials.

Process for Development of Oil and Gas Resources on Federal and Indian Lands Including Key Aspects of the Inspection and Enforcement Program

Development of oil and gas resources on federal and Indian lands is a multi-stage process. First, Interior holds auctions through which entities may secure the right to federal and Indian leases that allow them to drill for oil and gas after meeting certain conditions. Once an operator plans to drill a well on leased land, it must first secure a permit from Interior. After drilling a well, an operator installs production equipment, such as pump jacks, storage tanks, and metering equipment. This production phase continues until the well becomes inactive, and the operator may decide to plug the well, usually because the well is either depleted or no longer economically viable. After plugging the well, the operator is required to remove all production equipment and reshape and revegetate the land around the well.

To ensure compliance with applicable laws, regulations, and other requirements, BLM's Inspection and Enforcement program verifies that the operator complies with all requirements at a well or lease site during the drilling, production, and plugging phases.[15] Three BLM onshore orders, issued pursuant to regulation, specify requirements that operators are to follow on federal and Indian leases.[16] Inspectors use these orders to verify compliance during inspections.

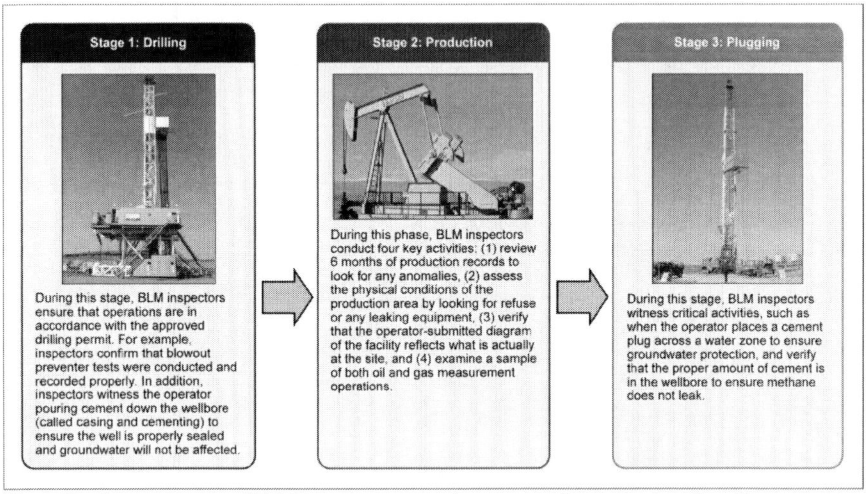

Note: During any stage, BLM inspectors can issue an enforcement action if an operator is found to not be in compliance with a requirement. Based on multiple factors, such as the severity of the violation, BLM enforcement actions can range from issuing a written order to correct the violation by a certain date, issuing a monetary assessment, or shutting down operations.

Figure 1. Key BLM Inspection Activities at Different Stages in the Life Cycle of a Well.

---

[15] The Federal Oil and Gas Royalty Management Act of 1982, as amended, Pub. L. No. 97–451, 96 Stat. 2447 (1983), codified at 30 U.S.C. §§ 1701, et seq., directs the Secretary of the Interior to develop guidelines that specify the coverage and frequency of inspections and authorizes the Secretary to enter any leases to make inspections, see 30 U.S.C. §§ 1711(a), 1718. Interior has delegated responsibilities for implementing the act to BLM for onshore leases.

[16] These three orders were first effective from March 27, 1989, through February 26, 1990, and in effect during the entire time period (i.e., fiscal years 2012 through 2016) covered in our audit. In November 2016, BLM issued new regulations that covered site security, oil measurement and gas measurement, which replaced these three onshore orders. See 43 C.F.R. subpart 3173, 43 C.F.R. subpart 3174, and 43 C.F.R. subpart 3175. These new regulations became effective on January 17, 2017, and, among other things, incorporated the latest industry standards.

Onshore Oil and Gas Order Number 3 specified requirements for the minimum standards for site security by ensuring that oil and gas produced from federal and Indian leases are properly handled to prevent theft and loss and enable accurate measurement. Onshore Oil and Gas Order Number 4 specified requirements for measurement of oil produced under the terms of federal and Indian leases or received by federal and Indian lessees as shares of oil produced on state or private lands. Onshore Oil and Gas Order Number 5 specified requirements for measurement of gas produced under the terms of federal and Indian leases or received by federal and Indian lessees as shares of gas produced on state or private lands.

Figure 1 shows key inspection activities that occur during the drilling, production, and plugging stages of a well's life cycle.

OUR ANALYSIS OF BLM DATA SHOWS THE DISTRIBUTION OF BLM'S INSPECTION AND ENFORCEMENT PROGRAM'S WORKLOAD AND WORKFORCE WAS OUT OF BALANCE IN FISCAL YEARS 2012 THROUGH 2016

In fiscal years 2012 through 2016, the distribution of the oil and gas Inspection and Enforcement program's workload and the workforce among the 33 BLM field offices with ongoing oil and gas development activities showed an imbalance, based on our analysis of BLM data. BLM took both short- and long-term actions in fiscal years 2012 through 2016 to address this imbalance, such as temporarily re-assigning inspectors from some medium activity offices to some of the highest activity offices.[17] Based on our review of BLM documentation and interviews with agency officials, two key factors affected the distribution of the program's workload: (1)

---

[17] Short-term actions were intended to provide immediate but temporary assistance to field offices in need of help, such as temporarily reassigning staff to those offices. Long-term actions were intended to provide long-lasting solutions to address the workload and workforce imbalance, such as permanently hiring additional staff.

energy market changes (e.g., price fluctuations) and (2) BLM actions to plan and prioritize inspection workload (e.g., changing risk classification for production inspections and decreasing the number of work months for plugging inspections).

The Distribution of the BLM Inspection and Enforcement Program's Workload and Workforce Was Out of Balance Based on Our Analysis of BLM's Data

From fiscal years 2012 through 2016, the distribution of the workload and workforce of BLM's oil and gas Inspection and Enforcement Program was out of balance across the 33 BLM field offices with ongoing oil and gas development activities, based on our analysis of BLM data. The majority of the workload, about 58 percent, was located at the 6 highest-activity field offices, which had 44 percent of the workforce.[18] In contrast, the majority of the workforce, 56 percent, was located in the remaining 27 medium and lowest activity offices, which had about 42 percent of the workload.

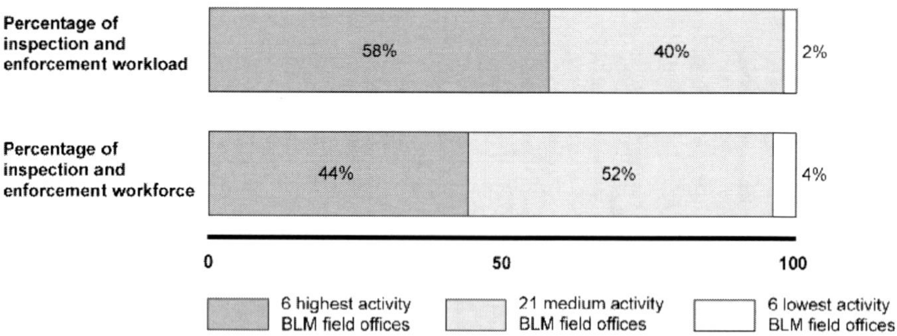

Source: GAO analysis of Bureau of Land Management (BLM) data. | GAO- 19-7.

Figure 2. Distribution of Workload and Workforce across the BLM Oil and Gas Inspection and Enforcement Program's 33 Field Offices, Fiscal Years 2012 through 2016.

---

[18] These six highest activity offices are Buffalo, Wyoming; Carlsbad, New Mexico; Dickinson, North Dakota; Farmington, New Mexico; Hobbs, New Mexico; and Tulsa, Oklahoma.

Figure 2 shows the distribution of workload and workforce across the 33 field offices. In addition, figure 3 shows a map of our categorization of BLM's 33 field offices by their workload and workforce activity level. From fiscal years 2012 through 2016, based on our review of BLM documentation and interviews with agency officials, BLM took both short and long-term actions to address this imbalance, such as temporarily re-assigning inspectors from some medium activity field offices to some highest activity offices.

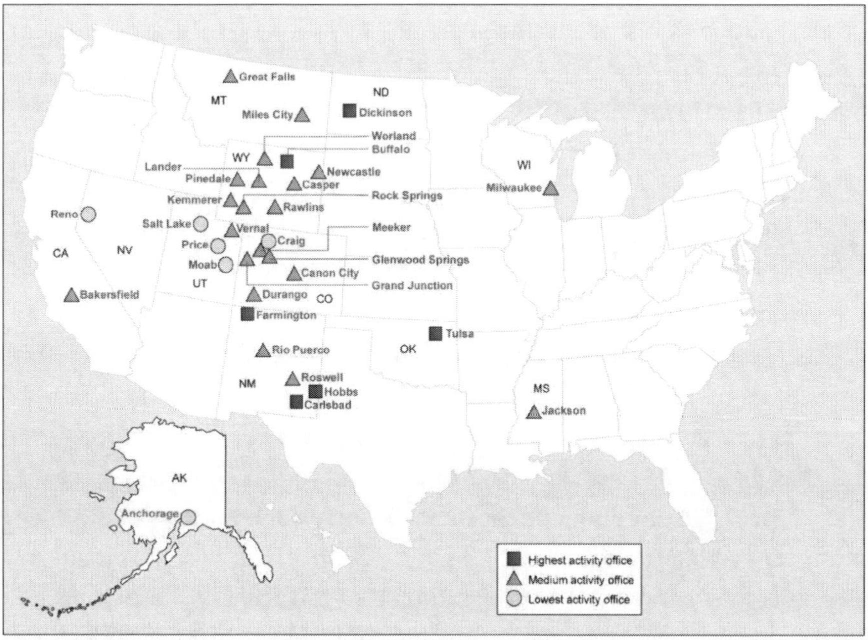

Source: GOA analysis of Bureau of Land Management (BLM) data; Map (map). | GAO-19-7.

Figure 3. BLM's Oil and Gas Inspection and Enforcement Program's Field Offices, by Workload and Workforce Activity Level, Fiscal Years 2012 through 2016.

A specific example of how BLM addressed this workload and workforce imbalance on a short term basis for this period concerns two of the highest activity offices (Hobbs and Dickinson). These offices had fewer PET inspectors on board and fewer PET inspection work months than three medium-activity offices (Pinedale, Rawlins, and Vernal). To address this imbalance, BLM sent short-term "strike teams" of PET inspectors to Hobbs and Dickinson on multiple occasions to help complete inspections. For example, officials from the Hobbs field office told us that in fiscal years 2012 and 2013, PET inspectors from the Farmington field office helped complete drilling and plugging inspections at Hobbs. In addition, officials from the Dickinson field office said that during fiscal years 2012, 2013, and 2014, more than 20 PET inspectors from five different states helped them inspect drilling, production, and plugging operations. BLM officials said there were pros and cons to the strike team approach. They said strike teams generally allow a field office to complete high-priority inspections and can provide additional training to inspectors at that office. However, agency officials said that, at times, the inspection documentation from strike team PET inspectors may not fully align with the policies and practices of the office they are assisting, which can create uncertainty about what inspection activities were completed and what the inspection found. We previously reported that strike teams increase costs and are not a sustainable solution.[19]

To address the workload and workforce imbalance on a long term basis, BLM allocated additional funding in fiscal years 2015 and 2016 to hire PET inspectors. The Inspection and Enforcement program manager said that these hires were targeted to address workforce needs at certain field offices. According to agency documentation, BLM allocated additional funding to hire about 20 inspectors in fiscal year 2015 and 40 inspectors in fiscal year 2016. Approximately 75 percent of these inspector positions were in three state offices: Montana (which includes the Dickinson, North Dakota field office), New Mexico (which includes the Tulsa, Oklahoma field office), and Wyoming. All six of BLM's highest activity field offices are located in these three states.

With this additional funding in fiscal years 2015 and 2016, multiple officials from BLM field offices reported that they were generally able to hire inspectors and, as a result, the number of onboard inspectors increased. For example, the number of onboard PET inspectors in the Dickinson field

---

[19] GAO-14-205.

office increased from 8 in fiscal year 2015 to 17 in fiscal year 2017. In the Buffalo field office, the number of onboard PET inspectors increased from 16 in fiscal year 2015 to 23 in fiscal year 2017. These officials generally cited two key reasons for being able to hire inspectors. First, BLM increased the compensation for PET inspectors through the use of special salary rates, incentive payments, and student loan repayments. We have previously reported that BLM faces challenges hiring PET inspectors because BLM competes with industry for employees, and industry offers higher salaries.[20] Second, and as described below, industry reduced development activity (i.e., wells drilled) in fiscal years 2015 and 2016 as commodity prices decreased. Multiple BLM field office officials also told us that it is easier to hire PET inspectors when oil and gas prices are low because industry is not hiring and applicants look to BLM for job security.

Two Key Factors Affected the Distribution of the Oil and Gas Inspection and Enforcement Program's Workload

Two key factors—based on our review of BLM documentation and interviews with agency officials—affected the distribution of the program's workload: (1) energy market changes (e.g., price fluctuations and increased development of shale plays) and (2) BLM actions to plan and prioritize inspection workload (e.g., changing risk classification for production inspections and decreasing the number of work months for plugging inspections). As we describe below, these factors affected several aspects of the program's workload (i.e., wells drilled, production inspection cases, planned plugging work months, and enforcement actions).

[20] GAO-14-205 and GAO, Oil and Gas Oversight: Interior Has Taken Steps to Address Staff Hiring, Retention, and Training but Needs a More Evaluative and Collaborative Approach, GAO-16-742 (Washington, D.C.: Sept. 29, 2016).

Consistently Lower Gas Prices, Volatile Oil Prices, and Increased Development of Shale Plays Led to a Decrease in Wells Drilled, but Not Uniformly Across BLM Field Offices

The number of wells drilled on federal and Indian lands from fiscal years 2012 through 2016 declined, according to BLM data. The decline was primarily the result of consistently lower gas prices and oil prices that dropped significantly in fiscal years 2015 and 2016 combined with technological advancements that increased the development of resources located in shale and other tight rock formations—which are generally not found on federal and Indian lands. Multiple BLM officials told us that commodity prices are a key factor that impacts the number of wells drilled on federal and Indian lands. These officials told us that, in general, when commodity prices are higher, industry will drill more wells, whereas when prices are lower, fewer wells are drilled. In addition, we previously reported that the highs and lows in prices and the number of oil and gas wells drilled largely overlapped, strongly suggesting that development activities reacted quickly and proportionally to changes in the prices of oil and gas.[21] Table 1 shows the number of wells drilled on federal and Indian lands and average monthly prices for natural gas and crude oil for the period. While there may have been some year-to-year variability between the number of wells drilled and commodity prices (see the fiscal year 2013 to 2015 prices for natural gas in table 1), operators drilled fewer wells in fiscal years 2015 and 2016, which were years of both consistently low gas prices and significant decreases in oil prices.

With regard to natural gas prices, a Purdue University study from March 2017 found that (1) the period of consistently lower natural gas prices (i.e., the Henry Hub average monthly price per million British thermal unit was generally from $2 to $4) began around 2009, which corresponds with increased development of natural gas from shale

---

[21] GAO, Onshore Oil and Gas: BLM's Management of Public Protests to Its Lease Sales Needs Improvement, GAO-10-670 (Washington, D.C.: July 30, 2010).

resources, and (2) the price increase in fiscal year 2014 was related to an extreme winter cold spell.[22]

Table 1. Number of Wells Drilled on Federal and Indian Lands and Average Monthly Prices for U.S. Benchmark Natural Gas and Crude Oil, Fiscal Years 2012 through 2016

Fiscal year	Number of oil and gas wells drilled on federal and Indian lands	Henry Hub natural gas average monthly price per million British thermal unit[a] (in dollars)	West Texas Intermediate crude oil average monthly price per barrel[b] (in dollars)
2012	3,521	2.73	95.25
2013	3,002	3.61	95.77
2014	3,180	4.40	98.71
2015	1,966	3.04	56.23
2016	998	2.28	41.20

Source: GAO analysis of Bureau of Land Management (BLM) data. | GAO-19-7.

[a] Henry Hub is a key hub located in Louisiana often used as a benchmark for United States natural gas.

[b] West Texas Intermediate is a light crude oil that is the most common benchmark in the United States.

With regard to oil prices, a World Bank report from January 2018 identified multiple factors contributing to the significant price decease that occurred in fiscal years 2015 and 2016.[23] These factors included increased oil production from U.S. shale plays—sedimentary rock formations containing significant amounts of oil and natural gas— contributing to oversupply as well as lower production costs that allowed shale oil wells to be profitable at lower prices.[24]

[22] Purdue University State Utility Forecasting Group, "Natural Gas Price Report Update," March 2017 (West Lafayette, Indiana).

[23] World Bank, "Global Economic Prospectus," January 2018 (Washington, D.C.).

[24] According to the Energy Information Administration, a play is a set of known or postulated oil and gas accumulations sharing similar geologic, geographic, and temporal properties, such as source rock, migration pathway, and hydrocarbon type. Oil and natural gas are found in a variety of geologic formations distributed across the country, such as shale or tight sandstone formations—also referred to as "tight oil" or "shale gas." Shale is a sedimentary rock that is predominantly composed of consolidated clay-sized particles.

From 2009 to 2016, there was also an increase in the development of oil and gas plays located in shale and other tight rock formations, brought about by advances in production technologies such as horizontal drilling and hydraulic fracturing.[25] According to Energy Information Administration data, shale plays represented more than 90 percent of the growth in oil and gas development from 2011 to 2016. As stated above, most shale plays are not located on federal and Indian lands.[26] However, the few BLM field offices located in shale plays where operators focus on oil development saw a smaller decrease in the number of wells drilled compared to field offices located outside of shale plays. For example, the Dickinson field office—located in the Bakken shale play—experienced a 15 percent decrease in the number of wells drilled from about 400 in fiscal year 2012 to about 330 in fiscal year 2016. Similarly, the Hobbs field office—located in the Permian shale play—experienced a 27 percent decrease from about 160 in fiscal year 2012 to about 120 in fiscal year 2016. According to BLM data, almost all producing wells in the Dickinson and Hobbs field offices are oil wells. In contrast, two field offices located outside of shale plays experienced a more significant decrease. The number of wells drilled in the Bakersfield field office (located in California) declined 90 percent from 285 wells drilled in fiscal year 2012 to 30 wells drilled in fiscal year 2016. According to BLM data, almost all of the Bakersfield field office's producing wells are oil wells. The number of wells drilled in the Vernal, field office (located in Utah) declined 95 percent from 725 wells drilled in fiscal year 2012 to 35 wells drilled in fiscal year 2016. According to BLM data, about 40 percent of the Vernal field office's producing wells are oil wells, and the remaining 60 percent are natural gas wells.

---

25 Hydraulic fracturing (also known as fracking) is commonly defined as an oil or gas well completion process that directs pressurized fluids to penetrate tight rock formations, such as shale formations, in order to stimulate and extract the oil or gas in the formation. The fluids typically contain a combination of water, proppant (a material that keeps an induced hydraulic fracture open), and added chemicals.

26 Specifically, we have previously reported that in 2016 about 15 percent of the major tight oil and shale gas plays in the contiguous United States overlapped federal lands, according to our analysis of Energy Information Administration and the U.S. Geological Survey data. GAO, Oil, Gas, and Coal Royalties: Raising Federal Rates Could Decrease Production on Federal Lands but Increase Federal Revenue. GAO-17-540 (Washington, D.C.: June 20, 2017).

BLM Changes to Risk Classification Led to Fluctuations in the Number of High Priority Production Inspection Cases

On multiple occasions from fiscal year 2012 through fiscal year 2016, based on our review of agency documentation, BLM changed its methodology to identify and classify risk, which led to fluctuations in the number of high-priority production inspection cases in a given fiscal year.[27] In our review, we focused on high priority production cases because, according to agency documents, inspecting such cases is one of the program's top three work priorities. Based on our review of agency documentation, BLM's risk-based strategy went through several iterations from fiscal years 2011 through 2016, and agency officials said that it was difficult to identify the specific reasons for year-to-year changes in the number of their high-priority production cases. This strategy used multiple weighted factors to develop a composite risk score to identify high- and low-priority cases. In fiscal year 2011, BLM based the composite risk score on seven weighted factors: four factors based on BLM data, and three factors based on data from Interior's Office of Natural Resources Revenue (ONRR).[28] However, BLM officials stated that they had challenges importing ONRR data in a format compatible with the bureau's information technology system and have since stopped using the data. From fiscal year 2013 through fiscal year 2016, BLM based the composite risk score on the following four BLM-identified risk factors: (1) average monthly production, (2) number of missing oil and gas operations reports, (3) number of incidents of noncompliance, and (4) number of years since last inspection. With regard to composite risk scores, in fiscal year 2011, BLM determined that a composite risk score of 4 would be considered high risk, meaning that cases with a score of 4 or more required an inspection. For fiscal year 2013, BLM increased the composite risk score needed to be considered high risk and required an inspection with a score of 5, a change intended to reduce the number of required inspections because agency

---

[27] As stated above, a case is either a lease or a unit agreement, can have from 1 to more than 1,000 wells, and is the unit of analysis for BLM's production inspection workload.

[28] ONRR is responsible for compiling data on the volume and value of leasable minerals produced from all federal lands and Indian lands where there is a trust responsibility, and collecting the appropriate payments.

documentation stated that the workload in the preceding years was too high for some field offices. For fiscal years 2014, 2015 and 2016, BLM lowered the composite risk score to 4 again.

BLM averaged about 2,150 high priority production cases in fiscal years 2012 through 2016, and in each of those fiscal years, the number ranged from about 1,700 to about 2,500. In addition, over 60 percent of such cases were located in the 6 highest-activity field offices we identified. Since such cases are concentrated in six field offices, seemingly minor fluctuations in the overall number of high priority production cases can have greater impacts to an individual field office's workload. For example, in fiscal year 2013, BLM identified about 2,500 high priority production cases. The Farmington field office in that year had about 170 such cases (or about 7 percent of the total) and estimated that PET inspectors needed about 12 work months to complete these inspections. In fiscal year 2015, BLM identified about 1,700 high priority production cases. The Farmington field office had about 90 such cases (or about 5 percent of the total) and estimated that PET inspectors needed about 6 work months to complete these inspections. In general, BLM officials told us that a single PET inspector is assigned about 6 inspection work months in a fiscal year once other demands on an inspector's time (i.e., sick leave, vacation, training, and the completion of other assigned non-inspection duties such as administering various safety programs) are considered. Therefore, in fiscal year 2013 the Farmington field office would have had to dedicate 2 PET inspectors (or about 10 percent of its total PET workforce) to complete only high priority production inspections, and in fiscal year 2015 the field office would have needed 1 PET inspector (or about 5 percent of its total PET workforce) to complete such inspections.

Since BLM's risk-based strategy has gone through multiple iterations since fiscal year 2012, several BLM officials said that it was difficult to identify the specific reasons for year-to-year changes in the number of their high-priority production cases. Officials, however, said that their ability to complete more high-priority production inspections increases during times of reduced industry drilling activity. Specifically, if industry is drilling fewer new wells, BLM can apply additional resources toward inspecting currently producing wells because PET inspectors who would normally conduct drilling inspections can now be deployed to high-priority production inspections. For example, as described above, the number of wells drilled decreased during the time frame covered in our review, with the Vernal and Bakersfield field offices experiencing substantial decreases in the number of wells drilled from fiscal year 2012 to fiscal year 2016. Officials in both offices told us that when drilling activity was low, BLM redirected resources originally planned for drilling inspections to complete high-priority production inspections.

BLM Reduced the Total Number of Estimated Well Plugging Inspection Work Months as Commodity Prices Stayed Low or Decreased

According to agency data, BLM reduced the estimated number of plugging inspection work months from about 200 in fiscal year 2012 to about 155 in fiscal year 2016, or about 23 percent.[29] Multiple agency officials told us that due to low or falling commodity prices operators plugged fewer wells from fiscal year 2012 through fiscal year 2016. As discussed above, natural gas prices were consistently low during fiscal years 2012 through 2016, while oil prices decreased significantly in fiscal years 2015 and 2016. According to multiple BLM officials, operators generally plug fewer wells during times of low or falling commodity prices because operators prefer to (1) maintain the income generated from even marginally producing wells or (2) limit the expenditures required to plug wells. In May 2018, we reported that low oil and gas prices placed financial stress on operators, increasing bankruptcies and the risk that operators would not permanently plug wells, and that BLM's actual costs and potential liabilities for reclaiming oil and gas wells likely increased for fiscal years 2010 through 2017.[30]

---

[29] At the start of each fiscal year, BLM estimates the number of work months it will need for all oil and gas activities that require inspections (e.g., drilling and well plugging). A work month is about 172 hours.

[30] GAO, *Oil and Gas Wells: Bureau of Land Management Needs to Improve Its Data and Oversight of Potential Liabilities*, GAO-18-250 (Washington, D.C.: May 16, 2018).

In addition, we reported that BLM faced challenges identifying and managing shut-in wells.[31] For example, BLM does not have time limits for how long operators can have a well in shut-in status, which may limit the agency's ability to ensure that operators permanently plug such wells before they become orphaned.[32]

However, since BLM estimates the number of plugging inspection work months at the start of each fiscal year, there can be instances where actual industry activity is different than estimated. For example, BLM officials at four field offices told us that during the time frame of our review, operators in their region plugged more wells than estimated. According to agency officials, these operators plugged more wells than BLM estimated because the operators were either looking to reduce their financial liability—sometimes in anticipation of selling assets—or looking for work to keep crews busy. In these instances, agency officials told us that, in general, BLM re-allocated inspection work months from low-priority production inspections to these plugging inspections. According to agency officials and documentation, plugging inspections are a higher priority than production inspections for multiple reasons. First, a plugging inspection is time sensitive because it is the final stage in a well's lifecycle. In contrast, a production inspection is an ongoing operation that can be conducted at almost any time. Second, properly plugging a well is essential for long-term environmental protection. For example, wells that are not properly plugged can leak methane and contaminate surface and groundwater. As such, multiple BLM officials told us that plugging inspections are their field office's highest priority work task and they will re-allocate resources, if necessary, to complete such inspections.

---

[31] A shut-in well is an inactive well that is not permanently plugged and that is physically and mechanically capable of producing oil or gas in paying quantities or capable of service use. For example, an operator may put a well in shut-in status if it has not been connected to a sales line or the line is too far away and it is not economical to connect to at this time.

[32] An orphan well is a well that generally has no responsible or liable party.

Higher Oil Prices in Some Years and Generally Lower Gas Prices Led to an Imbalance in the Program's Enforcement Workload at Three Field Offices

Based on our analysis of BLM data, two key market changes created an imbalance of the program's enforcement workload: (1) increased drilling activity at two field offices located in shale formations during times of higher oil prices, and (2) bankruptcies of coalbed methane operators in one field office as gas prices decreased. Combined, the Buffalo, Carlsbad, and Dickinson field offices issued about 45 percent of all enforcement actions, 75 percent of all monetary assessments, and about 85 percent of all civil penalties (see table 2). For purposes of this review, we focused on the number and amount of monetary assessments and civil penalties because, according to agency officials and BLM documentation, these two enforcement actions are the key tools used by BLM to address instances of serious or continued operator noncompliance.[33]

Table 2. Enforcement actions across BLM's Oil and Gas Inspection and Enforcement Field Offices, fiscal years 2012 through 2016

Field office	Number of written notices	Number of monetary assessments	Number of civil penalties	Number of shutdown of operations	Total number of enforcement actions
Buffalo, WY	12,648	711	370	223	13,952
Carlsbad, NM	5,167	271	26	1	5,465
Dickinson, ND	2,134	70	0	3	2,207
Remaining 30 field offices	26,797	349	79	60	27,285
Total	46,746	1,401	475	287	48,909

Source: GAO analysis of Bureau of Land Management (BLM) data. | GAO-19-7.

Notes: In general, BLM can issue enforcement actions of increasing severity—from written notices, to monetary assessments, to civil penalties, to shutdown of operations—to obtain operator compliance with federal requirements. Written notices are letters to operators that identify a violation, specify corrective action, and provide a compliance timeframe.

[33] Federal regulations also allow BLM to use other enforcement actions—such as bond forfeiture and lease cancellation—to address continued noncompliance issues. However, agency officials told us that these are rarely used for a variety of reasons, such as the lengthy administrative and judicial process required before a lease can be canceled.

Almost all of the monetary assessments that the Carlsbad and Dickinson field offices issued were for drilling violations—either drilling without approval or failure to install a blowout preventer or other well control equipment—and occurred in fiscal years 2012 through 2014, based on our review of BLM enforcement data. Federal regulations generally provide for higher monetary assessment amounts for drilling violations compared to other types of violations. Specifically, drilling violations are subject to assessments of $500 per day (up to $5,000), whereas a violation for failure to comply with a previously issued written notice for a minor violation is $250. As such, even though the Carlsbad and Dickinson field offices issued 24 percent of the number of monetary assessments, they issued about 60 percent (about $710,000) of the total amount assessed by all BLM field offices from fiscal years 2012 through 2016. In contrast, even though the Buffalo field office issued more than half of the monetary assessments, these actions accounted for 18 percent (about $220,000) of the total amount assessed because almost all of these assessments were minor violations for failure to comply (see table 3).

Table 3. Number, amount, and percentage of monetary assessments issued by BLM's Oil and Gas Inspection and Enforcement Field Offices, fiscal years 2012 through 2016

Field office	Number of monetary assessments	Percentage of assessments issued	Amount of monetary assessments (in dollars)	Percentage of assessed amount
Buffalo, WY	711	51	219,550	18
Carlsbad, NM	271	19	380,750	32
Dickinson, ND	70	5	327,000	28
Remaining 30 field offices	349	25	248,980	22
Total	1,401	100	1,176,280	100

Source: GAO analysis of Bureau of Land Management (BLM) data. | GAO-19-7.

From fiscal years 2012 through 2016, the Carlsbad and Dickinson field offices were responsible for about 30 percent of all wells drilled on federal and Indian lands, according to BLM data. These offices are located, respectively, in the Permian and Bakken shale plays, where almost all wells are oil wells. During fiscal years 2012 through 2014, for each of these field offices, operators drilled about 435 wells each year, and the price of oil ranged from $87 to $107 per barrel.[34] In contrast, during fiscal years 2015 and 2016, operators drilled about 275 wells each year while the price of oil ranged from $45 to $86 per barrel. According to agency officials, during fiscal years 2012 through 2014 operators attempted to drill wells as quickly as possible in the Carlsbad and Dickinson field offices to increase production during a time of higher oil prices.

BLM field office officials told us that when oil prices are higher, some operators have less financial incentive to follow federal requirements. In the Dickinson field office, for example, almost all monetary assessments were related to drilling without approval. Officials from that field office told us that, in general, these violations were related to operators who applied to BLM for a drilling permit, but the bureau did not approve the permit before the operator started drilling. In these instances, operators decided that the benefit of increased production at higher prices outweighed the cost of a monetary assessment, according to agency officials. BLM officials told us that for both types of drilling violations—drilling without approval and failure to install well control equipment—BLM issues monetary assessments immediately upon discovery due to the potential serious harmful impacts to resource development and environmental health and suspends drilling operations until the operator corrects the violation and pays the assessment. The officials said operators almost always pay these assessments in a timely manner because they wanted to complete drilling operations and start production.

In contrast to the monetary assessments issued during times of high oil prices, the Buffalo field office issued hundreds of civil penalties totaling millions of dollars during times of lower natural gas prices as some coalbed methane operators declared bankruptcy and did not

[34] As discussed earlier in this chapter, these prices are for West Texas Intermediate crude, the most common benchmark in the United States.

complete required reclamation activities.[35] Specifically, the Buffalo field office issued over 75 percent of the number of civil penalties and almost the entire amount penalized during fiscal years 2012 through 2016 (see table 4).

Table 4. Number, amount, and percentage of civil penalties issued by BLM's Oil and Gas Inspection and Enforcement Field Offices, fiscal years 2012 through 2016

Field office	Number of civil penalties	Percentage of penalties issued	Amount of civil penalties (in dollars)	Percentage of amount penalized
Buffalo, WY	370	78	16,273,931	97
Remaining 32 field offices	105	22	457,882	3
Total	475	100	16,731,813	100

Source: GAO analysis of Bureau of Land Management (BLM) data. | GAO-19-7.

As we reported in May 2018, low natural gas prices placed financial stress on operators of thousands of coalbed methane wells (natural gas extracted from coal beds).[36] In that May 2018 report, we also found that coalbed methane was economical to produce when natural gas prices were higher, and thousands of coalbed methane wells were drilled on federal lands.[37] However, coalbed methane production has declined because the production of shale gas has kept natural gas prices low. Officials from the Buffalo field office told us that (1) low natural gas prices contributed to an increasing number of bankruptcies among coalbed methane operators, and (2) in general, these bankrupt operators stopped production activities, shut-

[35] In general, reclamation activities include the removal of all production equipment, permanent plugging of the well and restoring lands to as close to their natural state as reasonably possible.

[36] GAO-18-250. In coalbed methane formations, gas is extracted through a process to reduce pressure called dewatering. As water is pumped out of the coal seams, reservoir pressure decreases, allowing the natural gas to release (desorb) from the surface of the coal and flow through natural fracture networks into the well. The separation of the gas from the water as well as the disposal of the water may be costly, making this process relatively expensive compared to natural gas production from formations containing less water.

[37] GAO-18-250.

in the wells instead of permanently plugging them, and stopped communicating with BLM.[38]

For these cases, Buffalo field office documentation outlines a 20-step process to identify a responsible party—that is, the operator or the person(s) to whom BLM issued the lease (the lessee)—to either permanently plug these wells or bring them back into production. Officials said that they repeated this 20-step process for each operator or lessee, as needed. Since one lease can have multiple lessees, the repetition of this process resulted in a very large number of enforcement actions, according to Buffalo field office officials. Under this process, BLM initially issued thousands of written notices requiring the responsible party to either "plug or produce." When the responsible party did not take the specified corrective action outlined in the written notices, the field office then issued hundreds of monetary assessments for failure to comply with the written notice and again instructed the operators to "plug or produce." When the responsible party failed to comply with the monetary assessments, Buffalo issued hundreds of civil penalties.[39]

Buffalo field office officials told us that they do not know whether the government has collected any of the issued penalties because the responsible parties did not pay the penalties to BLM in a timely manner. As such, BLM turned these outstanding penalties over to the Treasury Department for collection, a process that can take up to 2 years, according to agency documentation. Since market conditions have remained unfavorable for coalbed methane production, BLM has taken actions to

---

[38] A shut-in well is an inactive well that is not permanently plugged and that is physically and mechanically capable of producing oil or gas in paying quantities or capable of service use. If an operator plans to place a well into shut-in status for more than 30 days, BLM must approve the status change.

[39] In general, BLM oil and gas operations regulations authorize a fixed penalty amount or specify the maximum penalty amount (i.e., "up to") when BLM issues a civil penalty. According to agency documentation and the Inspection and Enforcement Program Manager, BLM's standard practice is to issue civil penalties at the maximum amount allowed by regulation. This practice provides consistency across BLM's 33 field office and does not put individual PET inspectors in the position to determine penalty amounts, according to the program manager. Federal regulations allow any operator who is issued a civil penalty to request an administrative review before the appropriate BLM State Director, who may reduce the amount assessed or penalized. Multiple BLM officials told us that these reviews rarely occur.

permanently plug some wells. For example, according to agency officials and documents, the agency has (1) worked with some non-bankrupt lessees, including at least one major oil and gas corporation, to plug wells, (2) re-directed funding from other BLM programs to pay to plug wells and (3) contributed funding to the state of Wyoming's well plugging program. We recently reported on BLM's actual costs and potential liabilities for reclaiming oil and gas wells and have ongoing work reviewing BLM's bonding requirements, which is the primary mechanism to ensure that operators complete required reclamation activities.[40]

BLM Has Not Completed All Required Internal Control Reviews of Its Field Offices and Does Not Employ a Risk-Informed Oversight Strategy

BLM state offices did not complete internal control reviews at 27 of 33 field offices—including 5 of the 6 highest activity offices we identified.[41] According to the July 2012 oversight policy, state offices are to periodically conduct internal control reviews of their field offices to, among other things, (1) review whether inspections and enforcement actions are accurate, complete, and conducted in accordance with policy, (2) review staffing and training needs, and (3) identify areas where program guidance can be improved. The July 2012 oversight policy also

---

[40] GAO-18-250.

[41] See U.S. Department of the Interior, Bureau of Land Management, *Oversight of the Oil and Gas Inspection and Enforcement Program*, Instruction Memorandum No. 2012-161 (Washington, D.C.: July 27, 2012). The July 2012 oversight policy also requires supervisors to conduct at least one oversight inspection on each inspector each fiscal year. Oversight inspections verify results of previous inspections and gauge inspection effectiveness, accuracy, and conformance to standards. From fiscal years 2013 through 2016, BLM conducted about 4,200 such oversight inspections, of which about 2,600 (62 percent) were conducted at the Farmington field office. Due to a large amount of missing required data—about 75 percent of all oversight inspections are missing the name of the inspector whose work was being reviewed—we could not confirm if each inspector received at least one oversight inspection each fiscal year, as required by the July 2012 oversight policy. BLM officials told us that this documentation issue should be addressed with enhanced data entry controls planned for its revised inspection record system, which agency officials said has a deployment goal of the second quarter of fiscal year 2019.

says that BLM state offices are responsible for overall programmatic oversight of field office operations. For those field offices with Inspection and Enforcement program functions, this means that state offices are responsible for ensuring that the field offices are able to meet the goals stated in the program's handbook, which include production accountability (i.e., the accurate measuring and reporting of production volumes), environmental safety, and public safety. BLM state offices completed internal control reviews at 6 of the 33 field offices from 2013 through 2017 and scheduled reviews for 5 others from 2018 through 2020, as shown in table 5.

Table 5. BLM Field Office Internal Control Reviews completed or scheduled since 2012

BLM state office	BLM field office	Reviews completed (Year)	Reviews scheduled(Year)
Alaska	Anchorage	-	-
California	Bakersfield	-	2019
Colorado	Canon City	-	-
	Craig	-	-
	Durango	2016	-
	Glenwood Springs	2013; 2017	-
	Grand Junction	2015	-
	White River/Meeker	2016	-
Eastern States	Jackson (MS)	-	-
	Milwaukee (WI)	-	-
Montana-Dakotas[a]	Dickinson (ND)	-	2019
	Great Falls	-	-
	Miles City	-	-
Nevada	Reno	-	-
New Mexico	Albuquerque	-	2020
	Carlsbad[b]	-	2018
	Farmington	-	2019
	Hobbs	2017	-
	Roswell	2016	-
	Tulsac (OK)	-	-

Table 5. (Continued)

BLM state office	BLM field office	Reviews completed (Year)	Reviews scheduled(Year)	
Utah	Moab		-	-
	Price		-	-
	Salt Lake		-	-
	Vernal		-	-
Wyoming	Buffalo		-	-
	Casper		-	-
	Rock Springs		-	-
	Kemmerer		-	-
	Lander		-	-
	Newcastle		-	-
	Pinedale		-	-
	Rawlins		-	-
	Worland		-	-
Totals	33		6	5

Source: GAO analysis of Bureau of Land Management (BLM) documents and interviews with officials. | GAO-19-7.

Legend:

- = indicates that the state office has either not completed or scheduled a review.

[a] The Montana-Dakotas state office also has jurisdiction over the Dickinson, North Dakota, field office. A Montana-Dakotas state official told us he had completed a review of the Great Falls field office but it was never formalized by management and he did not provide us a copy. The official also said reviews completed of the Miles City and Dickinson, North Dakota, field offices only focused on environmental inspections and not technical inspections and should not be included in our review.

[b] According to the New Mexico state inspection and enforcement coordinator, and as of November 2018, the internal control review of the Carlsbad field office was ongoing and scheduled to be completed by December 2018.

[c] The New Mexico state office also has jurisdiction over the Tulsa, Oklahoma, field office. A New Mexico state official told us the state office completed a review of the Tulsa field office in 2012, but neither the state official nor the BLM headquarters office was able to provide a copy of the review.

Officials from BLM state offices who completed internal control reviews said the benefits of these reviews included obtaining data to justify additional training or resources and providing a formal opportunity to examine key program management practices and correct identified deficiencies. For example, in September 2017, the Colorado state office completed an internal control review of a field office. Prior to this review, officials from that state office told us that they thought the field office might be understaffed based on a variety of factors, including longer than expected inspection times. BLM data showed that in fiscal year 2016 this field office estimated about 60 hours to complete a production inspection, while the other 5 Colorado field offices' average estimate was about 14 hours. The September 2017 Colorado state review identified unofficial management policies at this field office that resulted in the underutilization of PET inspectors and inflated inspection times, creating a perception of understaffing. For example, one of the field office's unofficial policies required that PET inspectors drive at least 1,000 miles a month in order to keep their government vehicle, which resulted in some inspectors taking longer routes and driving to locations beyond those required for the job. This policy contributed to artificially inflating inspection times. According to the Colorado state review, the accurate tracking of inspection times is vital for workload planning and staffing purposes. In response to these findings, the field office manager terminated the unofficial policy, and officials from the Colorado state office said they will check on the implementation of their recommendations by reviewing the inspection data. Officials from that state office are also no longer considering hiring additional PET inspectors. To ensure that the field office sustains these corrective actions, Colorado state officials told us that they perform periodic reviews of production inspection records and continue to hold progress report meetings with the field office's management team.

Although BLM state offices completed internal control reviews at 6 of 33 field offices, the state offices did not complete reviews at 27 field offices, including 5 of the 6 highest-activity field offices we identified. Officials from BLM state offices identified some key human capital and workload reasons that hindered their ability to complete reviews, including:

- long-term vacancies in multiple state offices' inspection and enforcement coordinator positions, which BLM filled on a temporary basis with other agency employees;

- competing priorities from upper management (e.g., preparing for lease sales); and
- hiring and training new PET inspectors.

For example, according to one state office inspection and enforcement coordinator, the coordinator position was filled on a temporary basis by four different BLM employees from about November 2013 to November 2015 as the agency tried to find a permanent hire. This official said that as a result of the personnel changes, the state office did not conduct field office internal control reviews as initially scheduled. In addition, another state office inspection and enforcement coordinator said that she spends a lot of her time providing instruction and on-the-job training to newly hired PET inspectors in multiple field offices that do not have a supervisory PET inspector, which limits her ability to perform field office internal control reviews.

We also identified two shortcomings with BLM's control activities that may have limited the agency's ability to compete internal control reviews as required by the July 2012 oversight policy. First, BLM headquarters did not design appropriate types of control activities to help management fulfill its responsibilities. Specifically, the Inspection and Enforcement program manager said that BLM headquarters did not consistently track and monitor the extent to which state offices completed field office internal control reviews. This official said that headquarters tends to rely on state offices to track and monitor such reviews and that headquarters focused on higher priority work tasks, such as developing and implementing new regulations that were issued in January 2017. Within the first 3 years following the issuance of the July 2012 policy, the agency completed one internal control review each during fiscal years 2013 and 2015, although at least 12 reviews were to be completed. BLM headquarters officials we spoke with were not aware that so few reviews had been completed in fiscal years 2013 and 2015.

Federal standards for internal control state that management should design control activities to achieve objectives and respond to risks, such as by comparing actual performance to planned or expected results and analyzing significant differences. Because it did not consistently monitor and track state office performance, BLM headquarters (1) did not know that state offices were not conducting field office internal control reviews in accordance with the July 2012 oversight policy and (2) could not analyze the reasons why actual performance did not meet expected results. Identifying the reasons it did not complete internal control reviews (e.g., human capital and workforce challenges), developing and implementing a plan to address those challenges, and monitoring state offices' progress toward completing required reviews will better position BLM to ensure that its state offices complete all required internal control reviews as called for by its July 2012 oversight policy.

Second, the July 2012 oversight policy identifies specific areas (e.g., the accuracy and completeness of inspections and staffing and training needs) that the reviews should assess, but according to a BLM headquarters official, the agency did not provide state offices with implementation guidance or procedures. This official said that BLM did not provide guidance or procedures so that state offices would have flexibility in how they conducted such reviews. However, multiple BLM state officials told us that such guidance or procedures would provide a helpful framework for conducting these reviews. One state office inspection and enforcement coordinator told us that since she had noprior training or experience designing and implementing internal control reviews, guidance or procedures would be especially beneficial.

Because they did not have documented implementation guidance or procedures to follow, the two state offices that completed internal control reviews developed their own procedures, which varied in design, methodology, and resources based our review of the six completed internal control reviews as well as interviews with BLM state officials. Specifically:

- One state office (1) developed its own review procedures based, in part, on existing program documentation, (2) assigned a single individual to conduct reviews because the state did not have the resources available to provide additional staff support, and (3) reviewed inspection and enforcement data contained in BLM's corporate oil and gas database as well as hard copy files, and interviewed field office PET inspectors.
- Another state office (1) developed its review procedures based on those employed during a 2011 review of the entire Inspection and Enforcement program at the suggestion of the Deputy State Director;[42] (2) assigned review teams consisting of multiple BLM officials with different areas of expertise; and (3) reviewed database and hard copy records, interviewed both field office PET inspectors and field office management, and observed field office PET inspectors as they conducted inspection activities.

Federal standards for internal control state that management should design control activities to achieve objectives and respond to risks, such as by clearly documenting internal control responsibilities in management directives, administrative policies, or operating manuals.[43] BLM has a documented policy, but this policy does not clearly specify what procedures state office officials are to follow to conduct internal control reviews. Without developing and documenting procedures for implementing internal control reviews under the July 2012 oversight policy, BLM does not have assurance that state offices will review all specific areas identified in the July 2012 oversight policy in a consistent manner.

---

[42] See United States Department of the Interior, Bureau of Land Management, 2011 Inspection and Enforcement Internal Control Review of Documentation of Inspections and Review of Drilling, Environmental, and Production Inspections (Attachment 1 to U.S. Department of the Interior, Bureau of Land Management, Internal Control Review of the Oil and Gas Inspection and Enforcement Program, Information Bulletin No. 2012-095, (Washington, D.C.: June 28, 2012)).

[43] GAO, *Standards for Internal Controls in the Federal Government*, GAO-14-704G (Washington, D.C.: Sept. 10, 2014).

In addition, although BLM did not have documented procedures for conducting periodic internal control reviews, the July 2012 oversight policy specified a schedule for conducting such reviews (see fig. 4). The schedule states the following:

- For state offices with four or fewer oil and gas field offices, the state office is to complete an internal control review of each field office at least once every 3 years. The state offices in this category are Alaska, California, Eastern States, Montana, Nevada, and Utah.
- For state offices with five or more oil and gas field offices, the state office is to complete an internal control review of each field office at least once every 6 years. The state offices in this category are Colorado, New Mexico, and Wyoming.

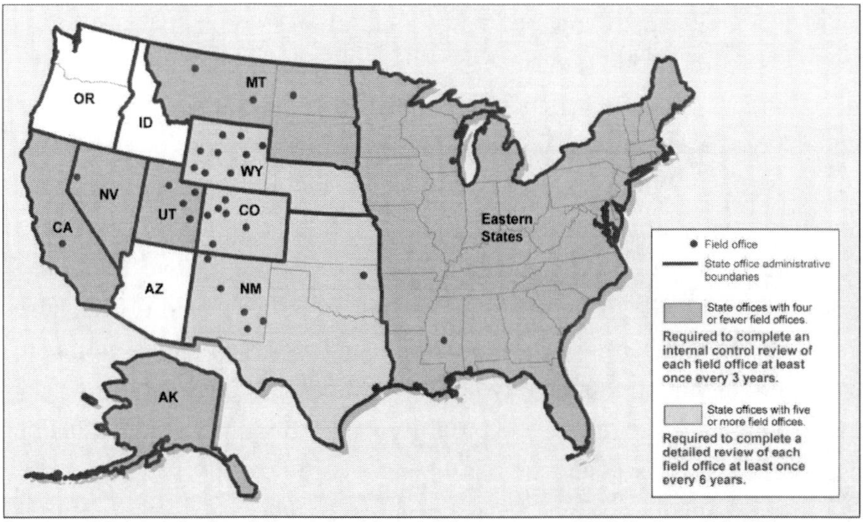

Source: GOA analysis of Bureau of Land Management (BLM) data; Map Resources (map). | GAO-19-7.

Figure 4. BLM Internal Control Review Schedule Determined by the Number of Field Offices within State Office Administrative Boundaries.

According to the Inspection and Enforcement program manager, this schedule was based on discussions with state office inspection and enforcement coordinators to balance officials' availability to conduct internal control reviews and other responsibilities. The program manager said that BLM did not identify or consider risk when developing the schedule because the agency's primary focus was to balance the new requirement to conduct field office internal control reviews with the state office coordinators' existing workload. However, the review schedule in the July 2012 oversight policy generally requires more frequent internal control reviews of low-activity offices and less frequent reviews of high activity offices. In particular, five of the six highest activity field offices we identified in our review are in states in which there are more than five field offices. According to the policy, these highest activity offices would therefore receive an internal control review at least once every 6 years. In contrast, five of the six lowest activity field offices are in states in which the policy requires that reviews be conducted at least once every 3 years. Such a review schedule may not ensure that BLM has properly established and implemented internal control reviews at the highest activity field offices—whose workforce must complete a majority of the program's workload—which may inherently pose a greater risk to the program's goals of production accountability, environmental protection, and personnel safety. For example, if the six highest activity field offices have an inadequate number of PET inspectors, then there is an increased risk to BLM's production accountability goal. Specifically, if these offices do not have the human resources needed to fully inspect high-priority production cases, BLM has less assurance that operators are properly measuring and reporting production volumes, which increases the risks to the accurate collection of royalty payments. Furthermore, those field offices that experienced greater levels of drilling workload may present a higher risk to BLM's environmental protection goal. Specifically, if the six highest activity offices do not conduct accurate and complete drilling inspections, BLM has less assurance that operators are properly conducting drilling operations, which increases the risks of environmental problems, such as contamination of fresh water aquifers.

Federal internal control standards call for entities to identify, analyze, and respond to risks related to achieving the defined objectives, such as by estimating the significance of identified risks to assess their effect on achieving defined objectives.[44] Management estimates the significance of a risk by considering the magnitude of impact, which refers to the likely magnitude of deficiency that could result from the risk and is affected by factors such as the size of a risk's impact. Without employing a risk-informed approach to scheduling and conducting internal control reviews that takes into account the risks to the Inspection and Enforcement program, such as those inherent in field offices' workload and workforce, BLM will not have reasonable assurance that it has adequate controls in place to address the effect of the field offices that pose the greatest risk to the program. BLM officials said that assessing risk, including field offices' workload activity levels, could provide a useful metric to inform how BLM conducts and prioritizes field office internal control reviews.

CONCLUSION

On federal and Indian lands, BLM's Inspection and Enforcement program is intended to ensure that operators developing oil and gas resources do so in a manner that protects public safety, environmental health, and royalty income. This is a complex undertaking that occurs within the oil and gas market and requires BLM's PET inspectors to conduct technically challenging drilling, production, and plugging inspections. In this context, BLM's July 2012 oversight policy calls for its state offices to conduct periodic internal control reviews of field offices. While BLM state offices completed internal control reviews at 6 field offices, they did not complete reviews at 27 field offices, including 5 of the 6 highest activity field offices we identified. In addition, because it did not consistently monitor and track state office performance, BLM headquarters (1) did not know that state offices were not conducting field office internal control reviews in accordance with the July 2012 oversight policy and (2)

[44] GAO-14-704G.

could not analyze the reasons why actual performance did not meet expected results. Identifying the reasons it did not complete internal control reviews (e.g., human capital and workload), developing and implementing a plan to address those challenges, and monitoring state offices' progress toward completing required reviews will better position BLM to ensure that its state offices are completing all required internal control reviews as called for by its July 2012 oversight policy.

Additionally, although BLM's July 2012 oversight policy does identify the specific areas that internal control reviews should assess, BLM did not provide state offices with implementation guidance or procedures. Because they did not have documented implementation guidance or procedures to follow, the two state offices that completed internal control reviews developed their own procedures, which varied in design, methodology, and resources. Without developing and documenting procedures for implementing internal control reviews under the July 2012 oversight policy, BLM does not have assurance that state offices will review all specific areas identified in the July 2012 oversight policy in a consistent manner.

Furthermore, and inconsistent with federal internal control standards, BLM's July 2012 oversight policy established a review schedule without identifying or considering risk. Without employing a risk-informed approach to scheduling and conducting internal control reviews that takes into account the risks to the Inspection and Enforcement program, such as those inherent in field offices' workload and workforce, BLM will not have reasonable assurance that it has adequate controls in place to address the effect of the field offices that pose the greatest risk to the program.

RECOMMENDATIONS FOR EXECUTIVE ACTION

We are making the following three recommendations to BLM:

The Director of BLM should identify the reasons internal control reviews were not completed (e.g., human capital and workforce), develop and implement a plan to address those reasons, and monitor state offices' progress toward completing required reviews. (Recommendation 1)

The Director of BLM should develop and document procedures for implementing internal control reviews under the July 2012 oversight policy. (Recommendation 2)

The Director of BLM should implement a risk-informed approach to scheduling and conducting internal control reviews that takes into account the risks to BLM's mission, such as those inherent in field offices' workload and workforce. (Recommendation 3)

AGENCY COMMENTS

We provided a draft of this product to the Department of Interior for comment. In its comments, reproduced in appendix I, Interior concurred with our three recommendations and outlined planned actions to implement them. For example, BLM plans to issue updated guidance and procedures for conducting internal control reviews to help ensure that such reviews are completed in a timely manner using a consistent risk-based approach.

As agreed with your office, unless you publicly announce the contents of this chapter earlier, we plan no further distribution until 30 days from the report date. At that time, we will send copies to the appropriate congressional committees, the Secretary of the Interior, and other interested parties.

Sincerely yours,

Frank Rusco

Director, Natural Resources and Environment

Appendix I: Comments from the Department of the Interior

United States Department of the Interior
OFFICE OF THE SECRETARY
Washington, DC 20240

FEB - 1 2019

Mr. Frank Rusco
Director, Natural Resources and Environment
U.S. Government Accountability Office
441 G Street NW
Washington, DC 20548

Dear Mr. Rusco:

Thank you for giving the Department of the Interior (Department) the opportunity to review and comment on the draft Government Accountability Office (GAO) report entitled, *Oil and Gas Development – Actions Needed to Improve Oversight of the Inspection and Enforcement Program* (GAO-19-7).

The GAO issued the Department's Bureau of Land Management (BLM) three recommendations in response to its overall findings. The Department agrees with the underlying findings leading to the three recommendations but proposes an alternative approach to resolve the GAO's findings.

The Department is committed to improving the oversight of BLM's oil and gas inspection and enforcement program. Additionally, the Department has identified the completion of 100% of the high priority fluid mineral case inspections during the year as an annual Strategic Plan goal in the new Fiscal Year (FY) 2018-2022 Strategic Plan.

The following is a summary of the BLM's planned actions to implement the report's recommendations:

Recommendation 1: The Director of BLM should identify the reasons internal control reviews were not completed (e.g. human capital and workforce), develop and implement a plan to address those reasons, and monitor state offices' progress toward completing required reviews.

Response: Concur. The BLM is aware that because of human capital (e.g., long-term vacancies in multiple state offices' inspection and enforcement coordinator positions) and competing priority challenges, some individual state-based internal control reviews were not completed. The BLM will implement an alternative solution by issuing updated guidance and procedures for conducting oversight reviews of the oil and gas inspection and enforcement program, including quality assurance. The guidance will ensure reviews are completed in a timely manner using a consistent risk-based approach.

Recommendation 2: The Director of BLM should develop and document procedures for implementing internal control reviews under the July 2012 oversight policy.

Response: Concur. The July 2012 oversight policy (IM-2012-161) required state offices to conduct internal control reviews with existing staff. This policy, which expired on September 30, 2013, should not have been allowed to lapse. Given the importance of internal control reviews, the BLM will issue updated guidance and procedures for conducting oversight reviews of the oil and gas inspection and enforcement program, including quality assurance.

Recommendation 3: The Director of BLM should implement a risk-informed approach to scheduling and conducting internal control reviews that takes into account the risks to BLM's mission, such as those inherent in field offices' workload and workforce.

Response: Concur. The BLM will issue updated guidance and procedures for conducting oversight reviews of the oil and gas inspection and enforcement program, including quality assurance. The guidance will ensure reviews are completed in a timely manner using a consistent risk-based approach.

Sincerely,

Joseph R. Balash
Assistant Secretary
Land and Minerals Management

In: Focus on Oil and Gas
Editor: Olivia B. Wilder

ISBN: 978-1-53616-325-4

Chapter 4

U.S. OIL AND NATURAL GAS TRANSFORMATION AND EFFECTS*

Michael Ratner, Laura B. Comay, Richard K. Lattanzio and Linda Tsang

OVERVIEW

U.S. oil and natural gas production has increased substantially since 2008 (Figure 1). These increases have important policy implications for energy markets, infrastructure, security, and the environment. These complex and interrelated implications have been both positive and negative, depending on perspectives, resulting in a variety of conflicts as well as enormous opportunities. Stakeholders, at times, turn to the courts to address conflicts, such as ensuring that fuel production and consumption comply with federal environmental laws.

---

* This is an edited, reformatted and augmented version of a Congressional Research Service publication IF11036, prepared for Members and Committees of Congress dated December 3, 2018.

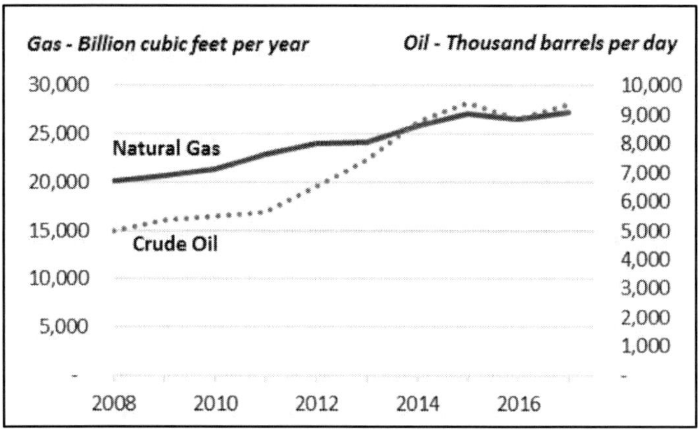

Source: Energy Information Administration (EIA). Prepared by CRS.

Figure 1. U.S. Natural Gas and Crude Oil Production.

In the 115th Congress, the House and Senate debated major energy legislation addressing expanding production, pipeline permitting, exports, and energy development on federal land, among other topics.

Oil and Natural Gas

The U.S. oil and natural gas industry has gone through a "renaissance" of production. Technological improvements in hydraulic fracturing and horizontal drilling, among other factors, have unlocked enormous oil and natural gas resources from unconventional formations, such as shale. Oil has surpassed levels of production not seen since the 1970s. Natural gas has set new production records almost every year since 2000. And the United States is the world's top producer of both commodities.

Commodity Prices

The expansion of natural gas supply since 2008 has led to a dramatic drop in prices, with implications for many different sectors, including electricity generation and manufacturing. Over that time, oil prices remained relatively volatile, but dropped in late 2014, and have remained lower since then. Note that while oil is traded on a global market, natural gas is much more of a regional commodity due primarily to transportation challenges. International price disparities for natural gas have diminished as more liquefied natural gas (LNG) has become available to global buyers.

Federal Lands

The rise in production of oil and natural gas has taken place mostly onshore and on nonfederal lands. Crude oil production from nonfederal land has doubled over the past decade. Although production on federal land has increased, it has not grown as fast as oil production on nonfederal land, causing the federal share of total (onshore and offshore) U.S. crude oil production to fall from its peak of nearly 36% in 2009 to about 24% in 2017. U.S. natural gas production shifted even more dramatically, with total U.S. production growing 33% since 2008, whereas gross withdrawals on federal lands (onshore and offshore) declined by almost 32% over the same time period. The federal share of total gross withdrawals decreased from 25% in 2008 to 13% in 2017. Through executive orders and subsequent implementing actions, the Trump Administration has set a federal lands agenda focused on U.S. energy dominance. Even with such a focus, any increases in production on federal lands may be outpaced by increases on nonfederal lands, particularly because the shale formations, where most of the growth is occurring, lie primarily on nonfederal lands.

Balancing energy production on federal lands against other resource values has long been a fundamental question for Congress. The 115th Congress debated this balance with respect to, among other issues, the Administration's proposed five-year program (2019-2024) for offshore oil and gas leasing; access to certain onshore federal lands for conventional or renewable energy development (including enactment of an oil and gas program for the Arctic National Wildlife Refuge); proposed changes to environmental and safety regulations; and efforts to streamline permitting. In court, stakeholder groups have challenged, and will likely continue to challenge, the federal leasing program for oil and gas development on federal lands. For example, some lawsuits seek to invalidate federal leases or leasing decisions or require the leasing agency to analyze risks to public health and the environment further.

Exports

As domestic production rose, industry sought to export more U.S. oil and natural gas (see Figure 2). The long-term prospects for further increases in either export remain unclear. With growing U.S. crude oil production and falling prices, there was interest in Congress in eliminating a 40-year limitation on exports of most U.S.-produced crude oil. In December 2015, Congress passed the Consolidated Appropriations Act for FY2016, which included a repeal of the oil export ban. Since the ban was lifted, exports of crude have reached over a million barrels per day or 10% of production. Additionally, U.S. exports of petroleum products, which were not limited by the law, have also increased significantly, almost tripling during the time period to 5.2 million barrels per day from almost zero. Total U.S. exports of 6.4 million barrels per day are still less than exports from Saudi Arabia.

The first U.S. LNG exports from the lower-48 states began in February 2016, but most LNG export projects remain in the construction or planning phases. Nevertheless, the United States became a net natural gas exporter in 2017, the first time in more than 50 years, mainly driven by increased pipeline exports to Mexico. Proposals have been introduced in the 115th Congress to expedite DOE's approval process.

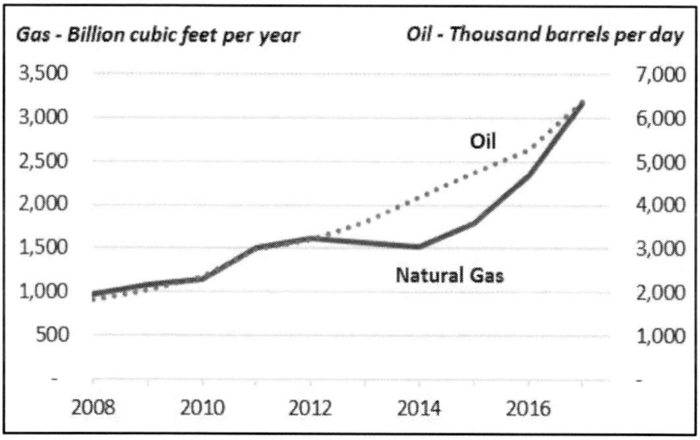

Source: Energy Information Administration (EIA). Prepared by CRS.

Notes: Natural gas exports include LNG and pipeline. Petroleum exports include crude oil and petroleum products.

Figure 2. U.S. Natural Gas and Petroleum Exports.

Infrastructure

Increased North American oil and gas production, particularly in areas that historically were not major producers, has led to growth in demand to transport those commodities to market. Oil shipments by rail and barge have increased significantly. Massive investments have been made in oil and gas pipelines, although controversy has arisen around projects such as the Dakota Access Pipeline and the Keystone XL Pipeline. As noted above, many projects aim to export LNG: these projects would encompass new terminal facilities at U.S. ports. In the 115th Congress, various bills would have amended rail safety standards, approved the Keystone XL pipeline, or promoted other oil and gas infrastructure. In addition, various legal challenges have tried to halt construction of pipelines and terminals to ensure adequate environmental review prior to construction.

Environmental Concerns

The use of unconventional oil and gas production has resulted in some significant environmental benefits (e.g., reduced air pollution from the substitution of natural gas for coal in power generation), but it has also raised concerns about other potential environmental and health effects. These concerns centered initially on water quality issues, including the potential contamination of groundwater and surface water from hydraulic fracturing and related production activities. Concerns have since incorporated other issues, such as water management practices (both consumption and discharge), land use changes, endangered species impacts, induced seismicity, and air pollution. Others have raised concerns about potential long-term and indirect impacts from reliance on fossil fuels and resulting greenhouse gas (GHG) emissions.

States are the primary regulators of oil and gas production on nonfederal lands, but various federal environmental statutes can apply to certain activities in the sector. (For example, the Clean Water Act regulates surface discharges of water associated with natural gas drilling and production as well as contaminated storm water runoff from production sites; the Safe Drinking Water Act regulates the underground injection of wastewater from crude oil and natural gas production and the underground injection of fluids used in hydraulic fracturing if the fluids contain diesel fuel; and the Clean Air Act limits emissions from associated engines and gas processing equipment as well as some natural gas extraction, production, and processing activities.) However, legislative proposals to address federal regulation have been highly controversial. Some advocates of a larger federal role point to a wide range of differences among state regulatory regimes and argue that a national framework is needed to ensure a consistent minimum level of protection. Others argue against more federal involvement and point to the long-established state regulatory programs, regional differences in geology and water resources, and concern over regulatory redundancy.

While congressional debate continues, the Trump Administration has pursued a number of deregulatory initiatives to promote oil and gas production. Executive Order 13783, "Promoting Energy Independence and Economic Growth" directs federal agencies to "review existing regulations that potentially burden the development or use of domestically produced energy resources and appropriately suspend, revise, or rescind" them. The executive order specifically cites several Obama-era rulemakings that regulated oil and gas production activities (e.g., the Environmental Protection Agency's methane standards and the Bureau of Land Management's waste prevention rule). Several states and other stakeholders have sought judicial review of the Trump Administration's efforts to delay, revise, or repeal existing environmental regulations.

STATES' ROLES

Many decisions about energy markets, infrastructure, and regulation are implemented at the state level. Regardless of whether the federal government takes a more or less active role in the future, state and regional decisions often have national impacts. For example, California's 2002 decision to regulate GHG emissions from automobiles has affected other states. The interaction of state and local policies with national decisions is often complex and can raise constitutional issues regarding the relationship between state and federal laws.

In: Focus on Oil and Gas
Editor: Olivia B. Wilder

ISBN: 978-1-53616-325-4

Chapter 5

GLOBAL OIL MARKETS AND U.S. GASOLINE PRICES 2018*

Phillip Brown and Robert Pirog

As the 2018 U.S. summer driving season begins, crude oil and gasoline prices have reached their highest levels since 2014. The international crude oil price benchmark (Brent) has increased by approximately 20% since the start of the year, with spot prices reaching just over $80 per barrel in mid-May. As a result, average U.S. retail gasoline prices— including taxes— have reached $2.99 per gallon (see Figure 1). Currently, crude oil contributes close to 60% of the price of gasoline. The remaining 40% is from refining, distribution and marketing, and taxes. Prices for other petroleum products such as diesel fuel and kerosene have risen as well.

Numerous factors can impact oil and petroleum product prices. Price drivers explored in this In Focus, namely market fundamentals (supply, demand, and stocks) and geopolitical risks, affect the nearly 100 million barrels per day (bpd) global market. However, other factors such as the

* This is an edited, reformatted and augmented version of Congressional Research Service, Publication No. IF10897, dated May 30, 2018.

strength of the U.S. dollar, global spare production capacity, and financial market expectations can also impact price levels.

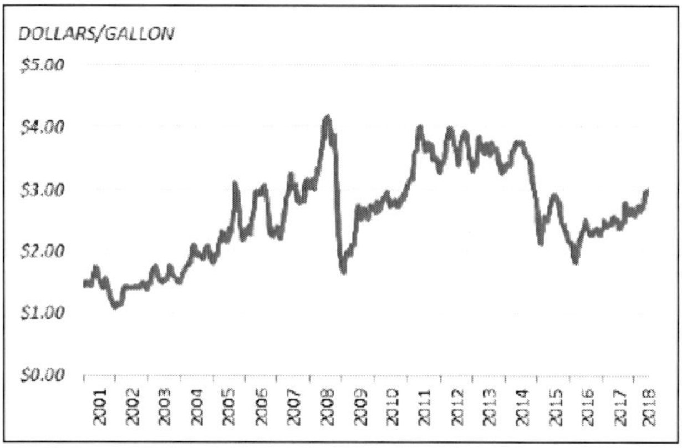

Source: Energy Information Administration.

Figure 1. U.S. Average Retail Gasoline Prices. Nominal Dollars.

Oil and Petroleum Product Market Fundamentals

Fundamental factors affecting the price of crude oil and petroleum products include global supply and demand, and the amount of, and changes to, global oil and petroleum product stocks. According to the U.S. Energy Information Administration (EIA), since the beginning of 2018 global supply has declined, demand has increased, and inventories have been subsequently reduced. Each of these changes to market fundamentals has resulted in upward price pressure that is reflected in current oil and gasoline prices.

Supply

On a quarterly basis, EIA reports that world oil and petroleum liquids production declined by approximately 150,000 bpd from the fourth quarter

Stocks

The difference between supply and demand is reflected in stock hanges. However, information about global stock levels is limited and tock levels for Organization for Economic Co-operation and Development OECD) countries is typically the metric used to gauge global stock hanges. One of the stated goals of the OPEC/Non-OPEC production greement was to reduce stock levels, which had reached record levels in 016, back to their five-year average. Since the OPEC/Non-OPEC roduction agreement took effect, OECD crude oil and petroleum product tock levels declined by 230 million barrels (8.3%). In February 2018, ECD stock levels were approximately 2.5 billion barrels (see Figure 2).

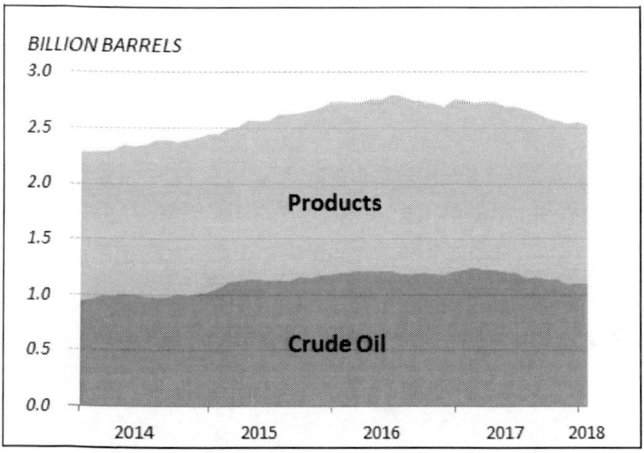

ource: International Energy Agency.

igure 2. OECD Crude Oil and Petroleum Product Stocks.

IEA reported in May that OECD stocks had reached the five-year verage. Based on this metric, the OPEC/NonOPEC agreement has guably achieved its objective. However, indications from OPEC embers suggest that the organization may change the metric used to etermine the success of the production agreement.

of 2017 through the first quarter of 2018. One oil market developme has been directly affecting global supply is an oil production agre between the Organization of the Petroleum Exporting Countries ((and several other oil producing countries. In November 2016, 11 then-active 13 OPEC members (Libya and Nigeria were e: announced an agreement to reduce oil production by approximat million bpd (Iran was allowed a small increase) from October 2016 The following month OPEC announced that 11 non-OPEC countri by Russia, had agreed to reduce oil production by an additional 5 bpd. This "Declaration of Cooperation" by 22 countries to redu production by approximately 1.8 million bpd came into effect in J 2017 and is currently scheduled to expire at the end of 201 agreement has been extended twice.

Group-level compliance with the OPEC/Non-OPEC agreem exceeded the commitment, and oil production by the parties ha reduced by approximately 1.9 million barrels per day since the agr took effect. A large portion of this "over-compliance" is unanticipated production declines in Venezuela (discussed further b

Counter-balancing the OPEC supply reduction to some degree, production has risen month-over-month by 1.4 million bpd b January 2017 and February 2018. Higher oil prices have benefited producers and are providing incentives to further increase productio

Demand

In its May report, the International Energy Agency (IEA) p 2018 global oil demand to increase by 1.4 million bpd compared While this was a downward revision of 100,000 bpd from tl report—attributed to anticipated demand declines from higher demand growth in 2018 is expected to be relatively high. This increase is driven primarily by the high growth rate of worldwi domestic product (GDP), which the International Monetary Fur estimates to be 3.9% for both 2018 and 2019.

Geopolitical Risk Factors

Global events with the potential to impact oil and petroleum product supply, demand, and trade can also affect price levels, especially in the short term when financial market sentiment can be influenced by risks and uncertainty. While developments in Syria, Yemen, and North Korea are significant factors, evolving situations in Venezuela and Iran, two major oil producers, could have a direct impact on short-term oil market conditions, including prices. Existing sanctions on Russia's energy sector will likely have limited short-term impacts and are therefore not discussed in this In Focus.

Venezuela

Venezuela is in the midst of a political and economic crisis that has contributed to a steep crude oil production decline. Since January 2016, crude oil production fell by nearly 1 million bpd month-over-month. Under the OPEC/NonOPEC agreement, Venezuela was to reduce production by 100,000 bpd from its October 2016 reference level of approximately 2.1 million bpd. In April 2018, the actual reduction compared to the reference level was 650,000 barrels per day. Further, the IEA projects that Venezuela production may decline through the end of 2018 by an additional 400,000 bpd. Should this projection materialize, there would likely be upward pressure on global oil prices. Additionally, the Trump Administration has indicated that it is considering petroleum sector sanctions on Venezuela. Depending on the structure of these potential sanctions, Venezuelan oil production could possibly decline further.

Iran

On May 8, 2018, President Trump announced that the Administration would reinstate all U.S. sanctions that were in place prior to the 2015 Joint

Comprehensive Plan of Action (JCPOA). For global oil markets, the most relevant of these sanctions are those aimed at reducing Iran's oil exports. The legal framework of these sanctions provides an incentive for oil buyers to "significantly reduce" purchases of Iranian oil. Between 2012 and 2015—when these sanctions were in effect—Iran crude oil production declined by approximately 1 million bpd. Iran oil production has since returned to pre-sanction levels of approximately 3.8 million bpd. Exactly how Iran crude oil production might be affected by the reinstatement of oil sector sanctions is uncertain. Current estimates suggest that Iran crude oil production could be reduced by as little as 100,000 bpd or as much as 1 million bpd. Nevertheless, should Iran oil production decline the resulting effect would likely be upward price pressure.

POLICY CONSIDERATIONS

During periods of escalating oil and petroleum product prices there is generally broad congressional interest in exploring policies that might provide some degree of price relief for U.S. consumers. In April, EIA reported that higher gasoline price projections would "result in the average U.S. household spending about $190 more on motor fuel in 2018." In May, the House Judiciary Committee held a hearing regarding the proposed No Oil Producing and Exporting Cartels Act (NOPEC). Subsequently introduced as H.R. 5904, the bill would amend the Sherman Act to make oil producing and exporting cartels illegal.

Additional bills have been introduced in the 115th Congress that can generally be categorized as efforts to address rising gasoline prices. S. 2929 would require the U.S. Trade Representative to pursue a complaint of anticompetitive practices against certain oil exporting countries. S. 2886 would reinstate restrictions on the export of crude oil and natural gas produced in the United States. Crude oil export restrictions were repealed in December 2015.

CONCLUSION

Generally, recent oil market trends and geopolitical factors have contributed to price escalation since the start of 2018. However, while current market fundamental data indicate demand has been exceeding supply, stock levels have declined, and prices have increased, current EIA projections for the remainder of 2018 and all of 2019 suggest that the situation may reverse and supply could potentially exceed demand. Two assumptions contributing to this supply surplus projection include continued U.S. oil production growth and unwinding the OPEC/Non-OPEC agreement when it expires at the end of 2018. EIA projections are subject to revision as market conditions change and geopolitical risks evolve. Should this occur, market fundamentals would be expected to exert downward pressure on oil and petroleum product prices.

In: Focus on Oil and Gas
Editor: Olivia B. Wilder
ISBN: 978-1-53616-325-4

Chapter 6

OFFSHORE OIL SPILLS: RESTORATION AND FEDERAL RESEARCH EFFORTS CONTINUE, BUT OPPORTUNITIES TO IMPROVE COORDINATION REMAIN*

United States Government Accountability Office

ABBREVIATIONS

BP	BP America Production Company
BSEE	Bureau of Safety and Environmental Enforcement
EPA	Environmental Protection Agency
NASA	National Aeronautics and Space Administration
NOAA	National Oceanic and Atmospheric Administration
NRDA	Natural Resource Damage Assessment
OPA	Oil Pollution Act of 1990

* This is an edited, reformatted and augmented version of United States Government Accountability Office; Report to the Chairman, Committee on Energy and Natural Resources, U.S. Senate, Publication No. GAO-19-31, dated January 3, 2019.

RESTORE Act	Resources and Ecosystems Sustainability, Tourist Opportunities, and Revived Economies of the Gulf Coast States Act of 2012
RESTORE Council	Gulf Coast Ecosystem Restoration Council
USGS	U.S. Geological Survey

Why GAO Did This Study

The *Exxon Valdez* and *Deepwater Horizon* oil spills are two of the largest offshore oil spills in U.S. history, causing long-lasting damage to marine and coastal resources. OPA includes provisions to prevent and respond to such oil spills by authorizing (1) federal-state trustee councils that manage billions of dollars from legal settlements and (2) an interagency committee to coordinate oil pollution research, among other things.

GAO was asked to review the federal government's response, restoration, and research efforts after the *Exxon Valdez* and *Deepwater Horizon* oil spills. This chapter examines, among other things, (1) how the trustee councils have used the restoration trust funds and the status of restoration and (2) the interagency committee's coordination of oil spill research efforts. GAO reviewed the councils' plans for the funds and how they were used, federal funding of oil spill research by member agencies, and key laws. Also, GAO evaluated the coordination of such efforts against a leading collaboration practice. GAO interviewed members of the trustee councils and the interagency committee.

What GAO Recommends

GAO recommends, among other things, that the interagency committee coordinate with the trustee councils to support their work and research needs. The agency agreed with GAO's recommendations.

WHAT GAO FOUND

The trustee councils, composed of federal and state members, have used portions of the restoration trust funds from the *Exxon Valdez* and *Deepwater Horizon* oil spill settlements to restore natural resources. From October 1992 to January 2018, the Exxon Valdez Oil Spill Trustee Council used about 86 percent of the fund's roughly $1 billion, primarily on habitat protection and restoration of damaged natural resources. According to the council, all but 5 of the 32 natural resources and human services identified as damaged by the spill have recovered or are recovering. The health of Pacific herring is one example of a resource that has not yet recovered. Further, the presence of lingering oil remains a concern almost 30 years after the spill. In May 2018, GAO accompanied trustee council researchers to the spill area and observed the excavation of three pits that revealed lingering oil roughly 6 inches below the surface of the beach, as captured in the photo below.

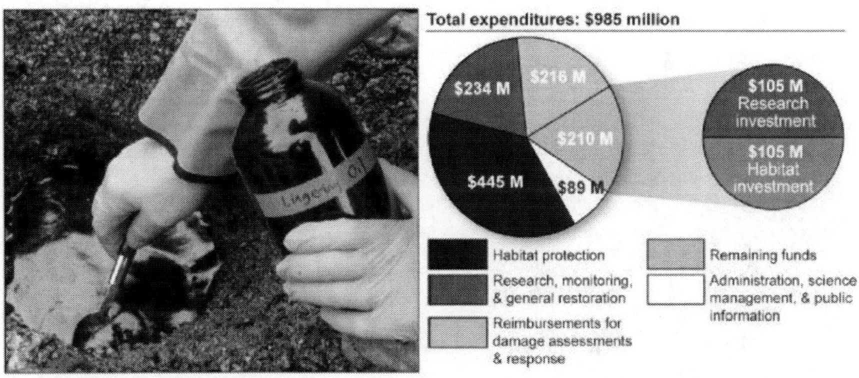

Source: GAO and GAO analysis of Exxon Valdez Oil Spill Trustee Council data. | GAO-19-31.

A Researcher Collects Lingering Oil from the *Exxon Valdez* Oil Spill on a Beach in Alaska in May 2018 (Left), and the Exxon Valdez Oil Spill Trustee Council Use of Restoration Trust Fund from October 1992 to January 2018 (Right).

The Deepwater Horizon Natural Resource Damage Assessment Trustee Council finalized a programmatic restoration plan in 2016; four trustee implementation groups have since issued initial restoration plans for designated restoration areas, and three anticipate issuing restoration plans in 2019 or later. From April 2012 to December 2017, the council used 13 percent of the at least $8.1 billion restoration trust fund, mostly on habitat protection, enhancing recreation, and marine wildlife and fishery restoration. The Oil Pollution Act of 1990 (OPA), which was enacted after the *Exxon Valdez* spill in 1989, established the Interagency Coordinating Committee on Oil Pollution Research (interagency committee) to coordinate oil pollution research among federal agencies and with relevant external entities, among other things. However according to the trustee council members that manage the restoration trust funds, the committee does not coordinate with the trustee councils and some were not aware that the interagency committee existed. The research of the member agencies could be relevant to the trustee councils' work on restoration. By coordinating directly with the trustee councils, the interagency committee could ensure better knowledge sharing between groups and leverage its member agencies' resources to inform and support the work of the councils.

January 3, 2019

The Honorable Lisa Murkowski
Chairman
Committee on Energy and Natural Resources
Unites States Senate

Dear Madam Chairman:

In March 1989, the supertanker *Exxon Valdez* ran aground in Alaska's Prince William Sound, spilling an estimated 11 million gallons of oil and contaminating about 1,500 miles of Alaska's south central coastline. At the time, this was the largest offshore oil spill in U.S. history. Enacted following the *Exxon Valdez* oil spill, the Oil Pollution Act of 1990 (OPA) included several provisions to prevent and respond to oil spills, such as requirements to assess damages to natural resources from oil discharges, including spills, and to develop and implement plans for restoration.[1] Approximately 20 years later, in April 2010, an explosion on the BP America Production Company's (BP) leased *Deepwater Horizon* oil rig spilled over 200 million gallons of oil into the Gulf of Mexico, setting a new record as the largest oil spill in U.S. waters.[2]

OPA established a "polluter pays" system, placing the primary burden of liability and costs of oil spills on the responsible party for the vessel or facility from which oil is discharged. Under this system, the responsible party assumes, up to a specified limit, the burden of paying for spill costs, including both removal costs (for cleaning up the spill) and damage claims (for restoring the environment and paying compensation to parties economically harmed by the spill). Above specified limits, a responsible party is no longer financially liable, although there are exceptions if, for example, the oil discharge is the result of gross negligence.

Following initial response and cleanup efforts, restoration activities related to a significant offshore oil spill, such as those from *Exxon Valdez* or *Deepwater Horizon*, can endure for decades. Under OPA, federal, state, tribal, and foreign government officials may be designated to act on behalf of the public as trustees of natural resources. In the wake of both oil spills, federal and state trustees entered into legal settlements with responsible parties to resolve certain claims. In October 1992, the federal government and the state of Alaska created the Exxon Valdez Oil Spill Trustee Council (Exxon Valdez Trustee Council) to administer the $900 million civil settlement designated for cleanup costs, damage assessment, litigation, and

[1] Pub. L. No. 101-380, 104 Stat. 484 (1990). As defined by regulation, restoration is any action (or alternative) or combination of actions (or alternatives) to restore, rehabilitate, replace, or acquire the equivalent of injured natural resources and services. 15 C.F.R. § 990.30.

[2] BP Exploration & Production, Inc. and BP America Production Company are direct or indirect wholly-owned subsidiaries of BP p.l.c. For purposes of this chapter, unless otherwise referring to specific subsidiaries or affiliates, we refer to "BP" as the BP entity responsible for payment under the civil settlement and criminal plea agreement stemming from the *Deepwater Horizon* incident.

restoration after the *Exxon Valdez* oil spill.[3] In April 2016, the federal government and affected states established a memorandum of understanding that created the Deepwater Horizon Natural Resource Damage Assessment Trustee Council (Deepwater Horizon Trustee Council) to administer the eventual $8.8 billion restoration fund[4] to address natural resource damages resulting from the *Deepwater Horizon* oil spill.[5]

OPA also established the Interagency Coordinating Committee on Oil Pollution Research (interagency committee) to coordinate a comprehensive program of oil pollution research[6], technology development, and demonstration among federal agencies in cooperation and coordination with external entities, such as industry, universities, research institutions, state governments, and other nations, as appropriate.[7] Federal agencies of the

---

[3] The members of the Exxon Valdez Trustee Council are the Attorney General of the state of Alaska, the Commissioner of the Alaska Department of Environmental Conservation, the Commissioner of the Alaska Department of Fish and Game, the Secretary of Agriculture, the Secretary of the Interior, and the Administrator of the National Oceanic and Atmospheric Administration (NOAA) in the Department of Commerce.

[4] This amount includes up to $700 million in potential additional payments for unknown conditions and adaptive management, intended to support future restoration and adaptive management by addressing conditions unknown at the time of the settlement. However, this money cannot be demanded before January 1, 2026; therefore, this chapter considers only the $8.1 billion minimum settlement amount when calculating the amount of funds used.

[5] The April 2016 memorandum of understanding superseded and replaced a November 2010 memorandum of understanding, which first created the Deepwater Horizon Trustee Council. The members of the Deepwater Horizon Trustee Council are the Department of the Interior, represented by the U.S. Fish and Wildlife Service, the Bureau of Land Management, and the National Park Service; the Department of Commerce, represented by NOAA; the Environmental Protection Agency (EPA); the Department of Agriculture; the Department of Environmental Protection and the Fish and Wildlife Conservation Commission of the state of Florida; the Department of Conservation and Natural Resources and the Geological Survey of the state of Alabama; the Department of Environmental Quality of the state of Mississippi; the Coastal Protection and Restoration Authority, Oil Spill Coordinator's Office, Department of Environmental Quality, Department of Wildlife and Fisheries, and Department of Natural Resources of the state of Louisiana; and the Commission on Environmental Quality, the General Land Office, and the Parks and Wildlife Department of the state of Texas.

[6] OPA established the interagency committee to coordinate a comprehensive program of oil pollution research. However, the interagency committee's research and technology plan refers to both oil pollution research and oil spill research. For the purpose of our report, we refer to oil pollution research when we describe OPA and the interagency committee's statutory responsibilities and to oil spill research in all other instances.

[7] Under OPA, the interagency committee is to include representatives from the U.S. Coast Guard in the Department of Homeland Security, the Department of Commerce (including the National Oceanic and Atmospheric Administration and the National Institute of Standards and Technology), the Department of Energy, the Department of the Interior (including the

interagency committee conduct and fund research projects related to preventing, preparing for, responding to, and restoring the environment after oil spills.[8] Later in this chapter, we describe in more detail the extent of research funded by the interagency committee. In accordance with OPA, the chair of the interagency committee—a representative of the U.S. Coast Guard (Coast Guard)—must submit a report to Congress every 2 years on its past and planned activities for oil pollution research. Under OPA, one of the interagency committee's responsibilities is to coordinate with federal agencies and external entities on an oil pollution research and development program that includes methods to restore and rehabilitate natural resources damaged by oil spills.

In March 2011, we reported that the interagency committee had taken limited actions to foster communication and coordination among member agencies and nonfederal stakeholders.[9] Among other things, we recommended that the Commandant of the Coast Guard direct the chair of the interagency committee, in coordination with member agencies, to provide a status update regarding the revisions of the committee's research plan and establish a more systematic process to identify and consult with key nonfederal stakeholders. The agency agreed with the recommendations, and the interagency committee implemented them by publishing an updated Oil Pollution Research and Technology Plan (research and technology plan) that includes a retrospective analysis of completed research as well as a forecast of new research needs.[10] In

---

Bureau of Safety and Environmental Enforcement, the Bureau of Ocean Energy Management, and the U.S. Fish and Wildlife Service), the Department of Transportation (including the Maritime Administration and the Pipeline and Hazardous Materials Safety Administration), the Department of Defense (including the Army Corps of Engineers and the Navy), the Department of Homeland Security (including the United States Fire Administration in the Federal Emergency Management Agency), the Environmental Protection Agency, the National Aeronautics and Space Administration, the United States Arctic Research Commission, and such other federal agencies the President may designate.

[8] Interagency committee members are not limited to funding research on one incident or spill. In contrast, the Natural Resource Damage Assessment trustee councils cannot fund basic research with restoration trust funds, but can fund research that directly contributes to restoring resources damaged by a specific spill.

[9] GAO, *Federal Oil and Gas: Interagency Committee Needs to Better Coordinate Research on Oil Pollution Prevention and Response*, GAO-11-319 (Washington, D.C.: Mar. 25, 2011).

[10] See Interagency Coordinating Committee on Oil Pollution Research, *Oil Pollution Research and Technology Plan Fiscal Years 2015-2021* (September 2015).

addition, the interagency committee has employed several sources of information to better understand ongoing research needs and activities within industry, academia, and the government.

To help inform offshore oil spill response efforts and decision making, government, industry, and academic scientists conduct research on the use and effect of various response techniques in different situations and environments, such as the icy waters of the Alaskan Arctic and the deep waters of the Gulf of Mexico. One such response technique used to manage the environmental impacts after the *Deepwater Horizon* oil spill was the application of chemical dispersants.[11] In May 2012, we reported on the use of dispersants and recommended that the Commandant of the Coast Guard direct the chair of the interagency committee in coordination with member agencies to, among other things, identify information on key dispersant research in its research and technology plans and periodically update and disseminate this information in its biennial reports to Congress.[12] We also recommended that the application of dispersants in subsurface and Arctic conditions be among the areas prioritized in future research. The member agencies generally concurred with our recommendations. In its research and technology plan for fiscal years 2015 through 2021, the interagency committee identified dispersants as one of the committee's standing research areas, specifically noting the use of dispersants in cold weather conditions and deep sea environments.

You asked us to review issues related to the federal government's response, restoration, and research efforts following the *Exxon Valdez* and *Deepwater Horizon* oil spills. This chapter examines (1) how the Natural Resource Damage Assessment (NRDA) trustee councils have used the restoration trust funds from the *Exxon Valdez* and *Deepwater Horizon* oil spills and the status of the restoration efforts, (2) the status of the interagency committee's oil spill research efforts and how coordination of such efforts has changed since we last reported on them in March 2011, and

---

[11] Dispersants help break down oil into small droplets that can more easily mix into the water below the surface, increasing biodegradation rates and potentially decreasing the impact of spilled oil on the shoreline.

[12] GAO, *Oil Dispersants: Additional Research Needed, Particularly on Subsurface and Arctic Applications*, GAO-12-585 (Washington, D.C.: May 30, 2012).

(3) what literature suggests about the effectiveness of various oil spill response techniques in the Arctic and the Gulf of Mexico.

To examine how the NRDA trustee councils used the restoration funds from the *Exxon Valdez* oil spill (from October 1992 to January 2018) and *Deepwater Horizon* oil spill (from April 2012 to December 2017) for restoration and the status of the restoration efforts, we obtained data from each trustee council on the amount of funds (1) ordered by the settlement for each restoration type; (2) authorized by the trustees for, but not yet spent on, restoration activities (authorizations); (3) spent on restoration activities (expenditures); and (4) not yet authorized for restoration activities (remaining balance) through calendar year 2017 for *Deepwater Horizon* and through January 31, 2018, for *Exxon Valdez*.[13] We assessed the reliability of the financial data that the NRDA trustee councils provided by, among other things, reviewing the annual reports for each trustee council and interviewing knowledgeable council staff about the steps they took to maintain these data and determined that the data were sufficiently reliable for the purposes of our report.[14] We also examined the approved restoration plans and, when available, annual reports on restoration activities, as well as reports and scientific studies that the trustee councils funded.[15] We met with officials from the Exxon Valdez Trustee Council to discuss the distribution of settlement money for restoration purposes after the *Exxon Valdez* oil spill, and with officials from the Deepwater Horizon Trustee Council, the Gulf Coast Ecosystem Restoration Council (RESTORE Council), and the National Fish and Wildlife Foundation to discuss the distribution of settlement money for restoration purposes after the *Deepwater Horizon* oil

[13] Our review focuses on the role of the NRDA trustee councils; therefore, we excluded the criminal fine and restitution payment for *Exxon Valdez* and the Resources and Ecosystems Sustainability, Tourist Opportunities, and Revived Economies of the Gulf Coast States Act of 2012 (RESTORE Act) funding and the criminal payments paid to the National Fish and Wildlife Foundation and other entities for *Deepwater Horizon* from the scope of this review.

[14] The Exxon Valdez Trustee Council has submitted all of its annual financial statements to a third-party audit. The Deepwater Horizon trustees are each responsible for procuring financial audits no less than once every 3 years, with the first audit occurring in 2017, according to the trustee council's standard operating procedures.

[15] Under OPA, damages for injuries to natural resources and services resulting from an oil spill incident must be based on a restoration plan developed after adequate public notice, opportunity for a hearing, and consideration of all public comments.

spill.[16] We also traveled to Alaska and the Gulf of Mexico to better understand the status of restoration.

To examine the status of federal oil spill research efforts by the interagency committee and how coordination of such efforts has changed since we last reported on it in March 2011, we requested funding data[17] and project information on oil spill research from all 15 member agencies of the interagency committee.[18] We received data from the nine member agencies that reported funding oil spill research projects from fiscal years 2011 through 2017.[19] These agencies provided data on agency expenditures on oil spill research and the research category of any projects funded. We assessed the reliability of these data by, among other things, reviewing related documentation and interviewing knowledgeable agency officials from the nine member agencies that provided data on the steps they took to maintain this information and determined that in most cases the data were sufficiently reliable for the purposes of our report.

We also interviewed officials at each of the 15 member agencies and compared their coordination practices to one of our federal leading practices for interagency collaboration to evaluate the interagency

---

[16] After the *Deepwater Horizon* oil spill, the RESTORE Act established a new trust fund for programs, projects, and activities that restore and protect the environment and economy of the Gulf Coast region. In addition to creating the trust fund, the RESTORE Act established the RESTORE Council, which includes the Governors of Alabama, Florida, Louisiana, Mississippi and Texas; the Secretaries of the Departments of Agriculture, the Army, Commerce, Homeland Security, and the Interior; and the Administrator of the Environmental Protection Agency.

[17] For this chapter we use "funded" to describe money spent of oil spill research projects. We asked member agencies to include the total number of research projects funded from fiscal years 2011 through 2017.

[18] As of November 2018, the 15 member agencies of the interagency committee are the Coast Guard, NOAA, EPA, the Bureau of Safety and Environmental Enforcement (BSEE), the U.S. Army Corps of Engineers, the Pipeline and Hazardous Materials Safety Administration, the U.S. Fish and Wildlife Service, the U.S. Navy, the National Aeronautics and Space Administration (NASA), the Department of Energy, the U.S. Fire Administration, the Maritime Administration, the National Institute of Standards and Technology, the Bureau of Ocean Energy Management, and the U.S. Arctic Research Commission.

[19] The nine member agencies that reported funding oil spill research from fiscal years 2011 through 2017 were the Bureau of Ocean Energy Management, BSEE, the Coast Guard, the Department of Energy, EPA, NASA, NOAA, the Pipeline and Hazardous Materials Safety Administration, and the U.S. Arctic Research Commission. The other six member agencies reported that they did not fund or conduct oil spill research from fiscal years 2011 through 2017.

committee's efforts to coordinate oil spill research.[20] We chose to focus on the collaboration practice pertaining to participants. In addition, we reviewed the 2013 interagency committee charter, the committee's most recent biennial reports to Congress covering fiscal years 2008 through 2017, and the committee's third multi-year research and technology plan for fiscal years 2015 through 2021. We also attended two committee meetings and reviewed minutes of eight past meetings. We reviewed OPA's provisions that established and govern the interagency committee's coordination efforts and membership, as well as various related executive documents.

To examine what literature suggests about the effectiveness of various oil spill response techniques in the Arctic and the Gulf of Mexico, we conducted a literature search for studies that analyzed and summarized the effectiveness of various oil spill response techniques in those regions. We identified existing literature from 1989 (the year of the *Exxon Valdez* oil spill) to March 2018 by conducting searches of various databases, such as Scopus and ProQuest. We chose to focus on three primary response techniques—mechanical recovery, in-situ burning, and the use of dispersants—used to clean up offshore oil spills. Our subject matter expert helped the team narrow the list of literature results and we relied on 16 studies and articles that we determined were relevant to our research objective. These studies and articles covered both regions and the three response techniques, but each study did not cover the entire range of these topics. For example, some of the studies were focused on one response technique in a specific region or environment. We supplemented the list of studies from these databases with information from our previous reports and those of the Congressional Research Service, the National Academies Press, the Environmental Protection Agency (EPA), the National Oceanic and Atmospheric Administration (NOAA), the Coast Guard, the American

---

[20] GAO, *Managing for Results: Key Considerations for Implementing Interagency Collaborative Mechanisms*, GAO-12-1022 (Washington, D.C.: Sept. 27, 2012). Because OPA requires coordination of the interagency committee, we use the term coordination to describe the committee's efforts and the term collaboration when referring to our prior work on collaboration leading practices. In our prior work, we use "collaboration" broadly to include interagency activities that others have variously defined as "cooperation," "coordination," "integration," or "networking."

Academy of Microbiology, and the Arctic Oil Spill Response Joint Industry Program. For a complete list of the literature, see the bibliography. We shared our summary of the literature search findings with agency officials representing some of the interagency committee members and included their perspectives where relevant. We did not independently evaluate the effectiveness of these response techniques.

For more details on our objectives, scope, and methodology, see appendix I.

We conducted this performance audit from July 2017 to January 2019 in accordance with generally accepted government auditing standards. Those standards require that we plan and perform the audit to obtain sufficient, appropriate evidence to provide a reasonable basis for our findings and conclusions based on our audit objectives. We believe that the evidence obtained provides a reasonable basis for our findings and conclusions based on our audit objectives.

BACKGROUND

OPA amended the Clean Water Act and established provisions expanding and consolidating the federal government's authority to prevent and respond to oil spills. This includes providing the federal government with the authority to perform cleanup immediately after a spill using federal resources, monitor the response efforts of the spiller, or direct the spiller's cleanup activities. OPA also established a "polluter pays" system, placing the primary burden of liability and costs of oil spills on the responsible party for the vessel or facility from which oil is discharged. Under this system, the responsible party assumes, up to a specified limit, the burden of paying for spill costs, including both removal costs (for cleaning up the spill) and damage claims (for restoring the environment and paying compensation to parties economically harmed by the spill).

OPA authorized the use of the Oil Spill Liability Trust Fund to fund up to $1 billion per spill incident for pollution removal costs and damages resulting from oil spills and mitigation of a substantial threat of an oil spill in navigable U.S. waters when a responsible party cannot or does not pay for the cleanup.[21] After the *Deepwater Horizon* oil spill, the Resources and Ecosystems Sustainability, Tourist Opportunities, and Revived Economies of the Gulf Coast States Act of 2012 (RESTORE Act) established a new trust fund for programs, projects, and activities that restore and protect the environment and economy of the Gulf Coast region as well as the RESTORE Council, which is to summarize its activities for each calendar year's activities in an annual report to Congress.

In addition, NOAA finalized regulations in 1996 for assessing natural resource damages resulting from a discharge or substantial threat of a discharge of oil.[22] The NRDA regulations recognize that OPA provides for designating federal, state, and tribal officials as natural resource trustees and authorizes them to make claims against the parties responsible for the injuries[23,24] Under NRDA regulations, a trustee council's work usually occurs in three steps: (1) a pre-assessment phase, (2) the restoration planning phase, and (3) the restoration implementation phase.[25] During the pre-assessment phase the trustees are to determine whether they have jurisdiction to pursue restoration. In the restoration planning phase the trustees are to evaluate information on potential injuries and use that information to determine the need for, type of, and scale of restoration. Finally, the restoration implementation phase describes the process for implementing restoration.

Under OPA, federal, state, tribal, and foreign government officials may be designated to act on behalf of the public as trustees of natural resources. In the wake of both the *Exxon Valdez* and *Deepwater Horizon* oil spills,

---

[21] The Coast Guard administers the fund through its National Pollution Funds Center. We previously examined the Coast Guard's management of the fund and made several recommendations to address uncertainty regarding future funding and the Coast Guard's internal controls for oil removal disbursements, among other things. GAO, *U.S. Coast Guard National Pollution Funds Center: Improved Controls Needed for Oil Removal Disbursements and Action Needed for Sustainable Funding*, GAO-15-682, (Washington, D.C.: Sept. 15, 2015).

[22] 61 Fed. Reg. 440 (Jan. 5, 1996). We refer to these regulations as the NRDA regulations.

[23] The NRDA regulations define injury as an observable or measurable adverse change in a natural resource or impairment of a natural resource service.

[24] 15 C.F.R. 990.11.

[25] 15 C.F.R. 990.12.

federal and state trustees entered into legal settlements with responsible parties to resolve certain claims. The Exxon Valdez Trustee Council is in the restoration implementation phase, while the Deepwater Horizon Trustee Council is in both the restoration planning and implementation phases. The National Oil and Hazardous Substances Pollution Contingency Plan, commonly known as the National Contingency Plan, contains the federal government's framework and operative requirements for preparing and responding to discharges of oil and releases of hazardous substances, pollutants, and contaminants. It establishes that federal oil spill response authority is determined by the location of the spill: the Coast Guard has response authority in the U.S. coastal zone, and EPA covers the inland zone.[26] In addition, NOAA is to provide scientific analysis and consultation during oil spill response activities in the coastal zones.

Exxon Valdez Oil Spill

The *Exxon Valdez* oil spill in Alaska's Prince William Sound in 1989 contaminated portions of national wildlife refuges, national and state parks, a national forest, and a state game sanctuary—killing or injuring thousands of sea birds, marine mammals, and fish and disrupting the ecosystem in its path. In October 1991, the U.S. District Court for the District of Alaska approved a civil settlement and criminal plea agreement among Exxon, the federal government, and the state of Alaska for recovery of natural resource damages resulting from the oil spill. Exxon agreed to pay $900

[26] The terms inland zone and coastal zone are defined in the National Contingency Plan (40 C.F.R. § 300.5). The coastal zone covers all waters subject to the tide; the Great Lakes; specified ports and harbors on inland rivers; waters of the contiguous zone; other waters of the high seas subject to the National Contingency Plan; and the land surface or land substrata, ground waters, and ambient air proximal to those waters. The inland zone refers to the environment inland of the coastal zone excluding the Great Lakes and specified ports and harbors on inland rivers. Spills in inland waters can potentially affect coastal waters and ecosystems, particularly if a spill occurs in water systems near the coast.

million in civil claims in 11 annual payments and $125 million to resolve various criminal charges.[27]

In August 1991, the federal government and the state of Alaska signed a memorandum of agreement and consent decree to act as co-trustees in collecting and using natural resource damage payments from the spill. The 1991 memorandum states that all decisions related to injury assessment, restoration activities, or other use of the natural resource damage payments are to be made by unanimous agreement of the trustees. According to the memorandum, the trustees are to use the natural resource damage payments to restore, replace, rehabilitate, enhance, or acquire the equivalent of the natural resources injured as a result of the oil spill and the reduced or lost services provided by such resources. The memorandum also recognized that EPA was designated to coordinate restoration activities on behalf of the federal government. In 1992, the trustees established the Exxon Valdez Trustee Council to ensure coordination and cooperation in restoring the natural resources injured, lost, or destroyed by the spill.

In 1994, the Exxon Valdez Trustee Council prepared a restoration plan for use of the funds, which consisted of five categories: (1) general restoration; (2) habitat protection and acquisition; (3) monitoring and research; (4) restoration reserve; and (5) public information, science management, and administration.[28] The restoration plan noted that in addition to restoring natural resources, funds may be used to restore reduced or lost services (including human uses) from injured natural resources, which includes subsistence, commercial fishing, recreation, and tourism services. The Exxon Valdez Trustee Council is advised by members of the public and a panel of scientists, and its Executive

---

[27] Of the $125 million, $25 million was a criminal fine and $100 million was restitution for the impact of the violations.

[28] Exxon Valdez Oil Spill Trustee Council, *Exxon Valdez Oil Spill Restoration Plan* (Anchorage, Alaska: November 1994). Under the terms of the consent decree for the settlement, in addition to providing for the restoration of injured natural resources, the $900 million civil settlement payments were to cover the costs for spill response and cleanup, damage assessment, and litigation. As of early 2018, the governments received $177 million as reimbursement for damage assessments and response, including cleanup and litigation, and Exxon received roughly $40 million as reimbursement for cleanup work it conducted in 1991 and 1992.

Director manages the day-to-day administrative functions. The Exxon Valdez Trustee Council has published documents that are on the council's public website, such as the Injured Resources and Services list (current as of 2014), lingering oil updates (current as of 2016), annual reports (current as of 2018), and annual project work plans (current as of 2018).

Deepwater Horizon Oil Spill

The *Deepwater Horizon* oil spill in the Gulf of Mexico in 2010 resulted in the tragic loss of 11 lives and a devastating environmental impact and affected the livelihoods of thousands of Gulf Coast citizens and businesses. In April 2016, BP, the federal government, and the five Gulf Coast states agreed to a settlement resolving multiple claims for federal civil penalties and natural resource damages related to the spill totaling up to $14.9 billion.[29] Under the terms of the consent decree for the settlement, BP must pay up to $8.8 billion in natural resource damages under OPA, which includes $1 billion BP previously committed to pay for early restoration projects, and up to $700 million to address injuries that were unknown to the trustees as of July 2, 2015, including for any associated Natural Resource Damage assessment and planning activities, or to adapt, enhance, supplement, or replace restoration projects or approaches that the trustees initially selected. BP is to make these payments into the Deepwater Horizon Oil Spill Natural Resource Damages Fund managed by the Department of the Interior (Interior), to be used jointly by the federal and state trustees of the Deepwater Horizon Trustee Council for restoration of injured or lost natural resources.

Two additional, separate restoration funds are to receive money from the BP civil and criminal penalties: (1) the Gulf Coast Restoration Trust Fund established under the RESTORE Act is to receive 80 percent of the

---

[29] In a separate, parallel agreement with the Gulf Coast states, BP agreed to pay up to $5.9 billion to resolve economic damages claims of state and local governments, bringing the total for all civil claims to $20.8 billion.

$5.5 billion Clean Water Act civil penalty paid by BP to support environmental restoration and economic recovery projects in the Gulf Coast region and (2) the Gulf Environmental Benefit Fund managed by the nonprofit National Fish and Wildlife Foundation is to receive $2.394 billion in criminal penalties. For more information on the amount and distribution of the BP civil and criminal payments, see Figure 1.

Prior to reaching the settlement in 2016, BP signed an agreement in April 2011 to provide $1 billion toward early restoration projects in the Gulf of Mexico to address injuries to natural resources caused by the spill. Early restoration projects may be developed prior to the completion of the injury assessment, which can take months or years to complete. Payments by BP for early restoration projects are counted towards its liability for the $8.8 billion in natural resource damages resulting from the spill.

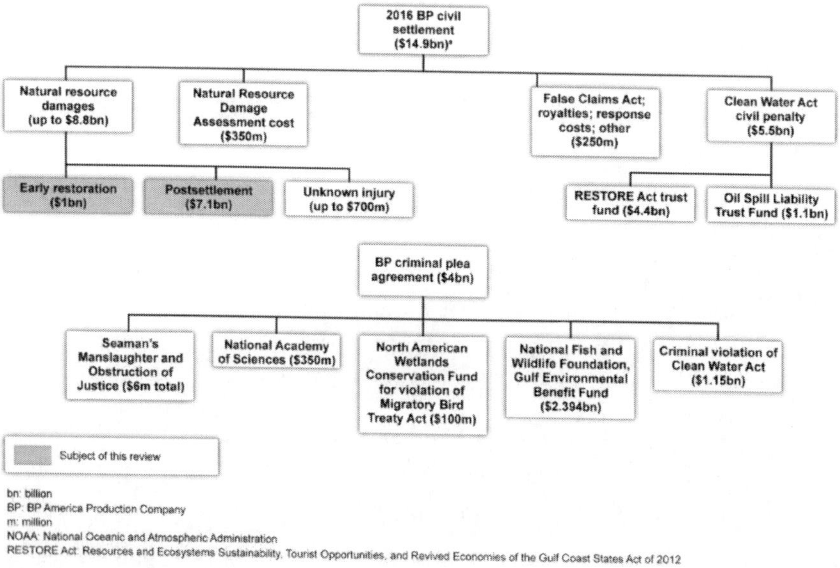

Source: GAO analysis of Department of Justice and RESTORE Council data. | GAO-19-31.

[a]There was a separate agreement for state and local economic claims; however, we excluded it from the figure because it was outside the scope of our review.

Figure 1. Distribution of BP's *Deepwater Horizon* Oil Spill Civil and Criminal Penalties.

The designated trustees are to administer these payments for natural resources, according to OPA. The designated trustees include federal officials from Interior, NOAA, the U.S. Department of Agriculture, and EPA, as well as state officials from the five Gulf States that were affected by the spill—Alabama, Florida, Louisiana, Mississippi, and Texas. In February 2016, the Deepwater Horizon Trustee Council finalized the Programmatic Damage Assessment and Restoration Plan (programmatic restoration plan) that provided the council's injury assessment and proposed a framework for identifying and developing project-specific restoration plans.[30] The five goals of the programmatic restoration plan are to (1) restore and conserve habitat; (2) restore water quality; (3) replenish and protect living coastal and marine resources; (4) provide and enhance recreational opportunities; and (5) provide for monitoring, adaptive management, and administrative oversight to support restoration implementation. According to the 2016 programmatic restoration plan, the Deepwater Horizon Trustee Council is to coordinate with other Deepwater Horizon restoration programs, such as those funded by the RESTORE Act, the National Fish and Wildlife Foundation, and other entities.

The 2016 programmatic restoration plan established Trustee Implementation Groups for each of the seven designated restoration areas—one for each of the five Gulf States, the Region-Wide implementation group, and the Open Ocean implementation group.[31] Each trustee implementation group is to plan, decide on, and implement restoration activities, including monitoring and adaptive management, for the funding that the consent decree allocated to its restoration area. Federal trustees serve in all the trustee implementation groups, and state trustees serve on the Region-Wide implementation group and the trustee implementation groups for their states; decisions are to be made by consensus. The Deepwater Horizon Trustee Council is to coordinate the

---

[30] Deepwater Horizon Natural Resource Damage Assessment Trustees, *Deepwater Horizon Oil Spill: Final Programmatic Damage Assessment and Restoration Plan and Final Programmatic Environmental Impact Statement* (Silver Spring, Maryland: Feb. 2016).

[31] In addition, the 2016 programmatic restoration plan established the Unknown Conditions and Adaptive Management implementation group, for which funding was set aside for use in the future.

work of the trustee implementation groups by establishing standard procedures and practices to ensure consistency in developing and implementing restoration activities.

Interagency Coordinating Committee on Oil Pollution Research

OPA created the interagency committee to provide a comprehensive, coordinated federal oil pollution research program and promote cooperation with industry, universities, research institutions, state governments, and other nations through information sharing, coordinated planning, and joint funding of projects. It also designated member agencies and authorized the President to designate other federal agencies as members of the interagency committee. As of November 2018, the interagency committee consisted of 15 federal members representing independent agencies, departments, and department components. OPA directs that a representative from the Coast Guard serve as the chair, and the interagency committee charter designates that a representative from NOAA, EPA, or the Bureau of Safety and Environmental Enforcement (BSEE) serve as the vice-chair and that the committee's Executive Director provide staff support. The interagency committee's charter notes that it shall meet at least semi-annually or at the decision of the chair. According to OPA, the chair's duties include reporting biennially to Congress on the interagency committee's activities related to oil pollution research, development, and demonstration programs.[32] OPA also required the interagency committee to prepare and submit a research and technology plan, which has been updated periodically.

In September 2015, the interagency committee released the research and technology plan for fiscal years 2015 through 2021. This research and technology plan updates the interagency committee's 1992 plan, revised in 1997, and provides a new baseline of the nation's oil pollution research needs. The plan is primarily directed at federal agencies with

[32] As of November 2018, the most recent biennial report to Congress by the interagency committee covers fiscal years 2016 and 2017 and was published on November 7, 2018.

responsibilities for conducting or funding such research, but it can also serve as a research planning guide for nonfederal stakeholders such as, industry, academia, state governments, research institutions, and other nations, according to interagency committee documents. The 2015 research and technology plan established a common language and planning framework to enable researchers and interested parties to identify and track research in four classes or categories that represent general groupings of oil spill research:

- Prevention: Research that supports developing practices and technologies designed to predict, reduce, or eliminate the likelihood of discharges or minimize the volume of oil discharges into the environment.
- Preparedness: Research that supports the activities, programs, and systems developed prior to an oil spill to improve the planning, decision-making, and management processes needed for responding to and recovering from oil spills.
- Response: Research that supports techniques and technologies that address the immediate and short-term effects of an oil spill and encompasses all activities involved in containing, cleaning up, treating, and disposing of oil to (1) maintain the safety of human life, (2) stabilize a situation to preclude further damage, and (3) minimize adverse environmental and socioeconomic effects.
- Injury assessment and restoration: Research that involves collecting and analyzing information to (1) evaluate the nature and extent of environmental, human health, and socioeconomic injuries resulting from an incident; (2) determine the actions needed to restore natural resources and their services to pre-spill conditions; and (3) make the environment and public whole after interim losses.

TRUSTEE COUNCILS USE RESTORATION TRUST FUNDS FOR APPROVED ACTIVITIES, WHICH ARE LARGELY COMPLETED FOR *EXXON VALDEZ* AND IN THE EARLY STAGES FOR *DEEPWATER HORIZON*

In response to the *Exxon Valdez* and *Deepwater Horizon* oil spills and by forming trustee councils, federal and state trustees have used the restoration trust funds to authorize money for activities in accordance with approved restoration plans. The Exxon Valdez Trustee Council has largely completed restoration work and authorized approximately $985 million, roughly 86 percent of the restoration trust fund, primarily on habitat protection and general restoration, research, and monitoring activities. As a result of these restoration activities and natural recovery, the majority of the injured natural resources and human services in the spill area has recovered or is recovering, according to the council's assessment. However, the Exxon Valdez Trustee Council continues to monitor the lack of recovery of Pacific herring and the presence of lingering oil in the spill area. The Deepwater Horizon Trustee Council is completing early restoration work and initial post-settlement restoration planning. It has authorized approximately $1.1 billion for restoration activities, roughly 13 percent of the restoration trust fund, and spent $368 million, roughly 5 percent of the restoration trust fund, primarily on habitat protection and enhancing recreation, such as building boat ramps and other recreational facilities.

The Exxon Valdez Trustee Council Has Used 86 Percent of the Restoration Trust Fund, and Most Injured Natural Resources Have Recovered

Exxon's payments to the restoration trust fund totaled approximately $900 million, and the interest earnings, as of January 2016, totaled $247 million. From 1992 to 2018, the Exxon Valdez Trustee Council authorized the expenditure of approximately $985 million or 86 percent of the roughly $1.15 billion in principal funds plus interest from the restoration trust fund, primarily on habitat protection ($445 million) and general restoration, research, and monitoring of injured natural resources ($234 million). The remaining unspent restoration trust fund balance as of January 2018 was $210 million, split evenly between the habitat investment subaccount for future habitat protection activities and the research investment subaccount for future general restoration activities (see Figure 2).

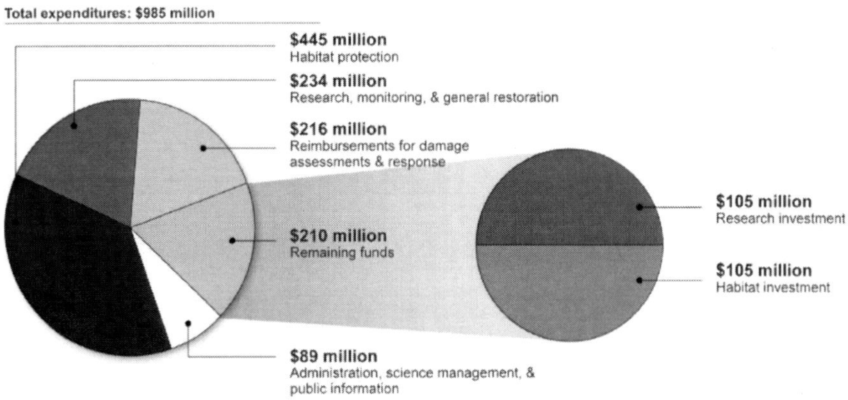

Source: GAO analysis of Exxon Valdez Spill Trustee Council data. | GAO-19-31.

Figure 2. Exxon Valdez Oil Spill Trustee Council Use of Restoration Trust Fund from October 1992 to January 2018.

According to the Exxon Valdez Trustee Council, as of January 2018, it had spent approximately $445 million to protect and enhance habitat, including acquiring 628,000 acres of lands and interest in lands.[33] As outlined in the trustee council's 1994 restoration plan, the habitat program is intended to minimize further injury to resources and services and allow recovery to continue with the least interference by authorizing funds for federal and state resource agencies to acquire title or conservation easements on ecologically valuable lands. For example, in 2017 the Exxon Valdez Trustee Council authorized about $5.5 million to acquire a

[33] The Exxon Valdez Trustee Council estimated that an addition $56 million was leveraged from other sources, such as state and federal agencies, to help acquire these lands.

conservation easement on 1,060 acres at the northeastern end of Kodiak Island in the Gulf of Alaska, known as Termination Point. The trustee council authorized funds for this acquisition to (1) protect the property from timber logging and development and (2) provide habitat and feeding areas for marine birds injured by the spill, such as marbled murrelets and pigeon guillemots. According to the Exxon Valdez Trustee Council, habitat acquisitions prevent additional injury to species during recovery, promote restoration of spill-affected resources and services, and are the primary tool for acquiring equivalent resources harmed by the spill. The habitat program also supports habitat enhancement projects, which, according to the Exxon Valdez Trustee Council, aim to repair human-caused harm to natural resources, their habitats, and the services they provide to humans. For example, the trustee council authorized $2.2 million to the Alaska Department of Natural Resources to stabilize stream bank vegetation and install elevated steel walkways to provide less-damaging access to the Kenai River, a popular fishing destination.

The Exxon Valdez Trustee Council has spent roughly $234 million from October 1992 to January 2018 on hundreds of general restoration, monitoring, and research activities. As outlined in the 1994 restoration plan, general restoration includes activities that manipulate the environment, manage human use, and reduce marine pollution. Research and monitoring activities also provide information on the status and condition of resources and services, including (1) whether they are recovering, (2) whether restoration activities are successful, and (3) factors that may be constraining recovery, according to the 1994 plan. For example, since 2012, the trustee council has authorized money for a program called Gulf Watch Alaska that provides long-term monitoring data on the status of environmental conditions—such as waters temperature and salinity—and the marine and nearshore ecosystems. Gulf Watch Alaska provides data to federal, state, and tribal agencies, as well as the public, that informs resource conservation programs and aid in the management of species injured by the spill. According to the trustee council, its expenditures for research projects have resulted in hundreds of peer-reviewed scientific studies and increased knowledge about the marine environment that benefits the injured resources.

The Exxon Valdez Trustee Council has spent roughly $89 million from October 1992 to January 2018 on administration, science management, and public information. According to the 1994 restoration plan, expenditures under this category cover the cost to (1) prepare work plans, (2) negotiate habitat purchases, (3) provide independent scientific review, (4) involve the public, and (5) operate the restoration program. Although the Exxon Valdez Trustee Council set a target of 5 percent administrative costs in the 1994 restoration plan, according to a written statement that the trustee council provided, administrative costs averaged around 6 percent from 1994 through 2001. The trustees and council staff we interviewed told us that in hindsight the 5 percent target was unrealistic as it did not reflect the actual administrative costs at that time, although such costs were included in project budgets or were absorbed by federal and state agencies. Therefore, in 2012, the Exxon Valdez Trustee Council changed the way it accounted for administrative costs and has included these costs in the administrative budget. According to the trustee council, under the new accounting policy, administrative costs were recalculated and estimated at around 19 percent for the period 2002 through 2018.

The remaining $210 million Exxon Valdez restoration trust fund balance is held by the Alaska Department of Revenue in two interest-bearing subaccounts.[34] As of January 2018, the research subaccount and the habitat subaccount each held approximately $105 million.[35] In the 1994 restoration plan, the Exxon Valdez Trustee Council established the need for a restoration reserve to ensure that restoration activities could continue to be supported after the final annual payments from the Exxon Corporation were received in September 2001. According to the 1994

[34] These funds are invested in the stock market, and interest earned is based on market returns. According to the Exxon Valdez Trustee Council, these funds have been invested with a fairly aggressive asset allocation and have had historic market returns in recent years; however, over the coming years the investment portfolio will shift to a more conservative one with a lower-risk profile to allow for consistent funding flows in the fund's final years.

[35] According to the Exxon Valdez Trustee Council, additional funds have been obligated since January 2018 but not yet spent, and taking these amounts into account, the total remaining funds from both accounts would be approximately $169 million as of June 2018.

restoration plan, the trustee council planned to set aside $12 million per year for a period of 9 years into the restoration reserve, totaling $108 million plus interest. In 1999, the Exxon Valdez Trustee Council resolved to transfer the estimated remaining balance of $170 million to the restoration reserve and split the money into two subaccounts. Since 2002, the trustee council is to make allocations for its annual work plans and ongoing habitat acquisition using these accounts.

In 2010, the trustee council established a 20-year strategic plan to spend the remaining trust funds using four 5-year incremental work plans. In November 2010, the trustee council issued a call for project proposals for the first 5-year work plan, for fiscal years 2012 through 2016. Although the Exxon Valdez Trustee Council solicited invitations on a 5-year cycle, it has authorized money for each project annually. In a written statement, the trustee council also stated that it continues to pursue and acquire from willing sellers remaining parcels of land that prior studies have identified as high-priority habitat. According to the Exxon Valdez Trustee Council's long-term spending scenario, both of the subaccounts are expected to be depleted by 2032 or earlier as determined by the market's performance.

The Status of Restoration Efforts

According to the Exxon Valdez Trustee Council's 2014 restoration plan update—its most recent assessment of injured resources and services— all but 5 of the 32 natural resources and human services identified as injured by the spill have recovered, are recovering, or are very likely recovered.[36] In the 1994 restoration plan, the trustee council

[36] Exxon Valdez Oil Spill Trustee Council, *Exxon Valdez Oil Spill Restoration Plan: 2014 Update, Injured Resources and Services* (Anchorage, AK: November 2014). This document provides an update on the status of injured resources and services that were first identified in the 1994 restoration plan, called the Injured Resources and Services List. In 2010, the Exxon Valdez Trustee Council introduced a fifth recovery status (very likely recovered) to reflect the status of species for which (1) there has been limited scientific research on the resource's recovery status in recent years; (2) prior studies suggest that there had been substantial progress toward recovery in the decade following the spill; and (3) so much time has passed since any

established a list of resources and services that suffered injuries from the spill, and developed specific, measurable recovery objectives for each injured resource and service.[37] The Exxon Valdez Trustee Council has periodically assessed the status of those resources, most recently in 2014. As of the 2014 assessment, the following 4 resources were listed as not recovering: (1) marbled murrelets, (2) Pacific herring, (3) pigeon guillemots, and (4) one group of killer whales. In addition, the recovery of Kittlitz's murrelets was listed as unknown.[38] According to the Exxon Valdez Trustee Council, the status of these resources in 2018 is largely similar to their status in 2014 except that one population of pigeon guillemots has likely increased as a result of a predator-control project that the council supported. However, the overall status of this species has not been determined. In a written statement, the trustees stated that the trustee council plans to initiate its next assessment of injured resources in late 2018.

The Exxon Valdez Trustee Council remains particularly concerned about the health of the Pacific herring population and the presence of lingering oil. According to the trustee council's 2014 restoration plan update, Pacific herring are considered an ecologically and commercially important species that in addition to being fished for human consumption is a source of food for various marine species. The assessment noted a combination of factors, including disease, predation, and poor recruitment of additional fish to the stock through growth or migration, appear to have contributed to the continued suppression of herring populations. As a result, the herring fishery has been closed for 23 of the 29 years since the oil spill and has not met the trustee council's recovery objective. To

---

indications of some spill injury, including exposure to oil, that it is unlikely that there are any residual effects of the spill.

[37] The Injured Resources and Services List does not necessarily contain all species affected by the spill, and the Exxon Valdez Trustee Council may support a species or habitat beyond its prespill status in part to make up for species damaged by the spill that will not likely recover, according to the Exxon Valdez Trustee Council.

[38] According to the assessment, for resources in the unknown category, data on life history or the extent of injury from the spill are limited. Moreover, given the length of time since the spill, it is unlikely that new or further research will provide information that will help in comprehensively assessing the original injury or determining the residual effects of the spill such that a better evaluation of recovery can occur.

address concerns regarding the Pacific herring, the trustee council plans to authorize additional money for ongoing Pacific herring research and monitoring through the anticipated end date for the fund in fiscal year 2032, for an estimated total cost of roughly $23 million over 20 years.

The Exxon Valdez Trustee Council also has concerns regarding the presence of lingering oil in the spill area. According to a March 2016 report for the trustee council, approximately 27,000 gallons of lightly weathered oil from the *Exxon Valdez* spill remains, located along almost 22 miles of shoreline at a small number of subsurface sites, where oxygen and nutrients are at levels too low to support microbial degradation.[39] In May 2018, we accompanied researchers working with the trustee council to the spill area and observed the excavation of three pits that revealed lingering oil roughly 6 inches below the surface of the beach, as captured in Figure 3. According to the researchers, oil previously recovered from this location was identified as belonging to the *Exxon Valdez* oil spill. Evidence of exposure to lingering oil was observed as recently as 2009 in a variety of marine species, including sea otters and harlequin ducks, according to the 2016 lingering oil report. The report also noted that the most recent studies show that the sea otter and harlequin duck populations have recovered and that lingering oil is no longer causing ecological damage. Further, studies demonstrated that minimally intrusive remediation of the oil would only be effective at a small number of sites, according to the 2016 report. Therefore, although the trustee council has decided not to pursue remediation of the oil, it stated that it has authorized money for projects to study the effects of oil and lingering oil totaling over $16 million and will continue to monitor the oil to document its physical and chemical changes over time. The Exxon Valdez Trustee Council expects that lingering oil

[39] Research Planning, Inc. and U.S. Geological Survey, *Studies on Exxon Valdez Lingering Oil: Review and Update on Recent Findings* (Anchorage, Alaska: March 2016). This chapter summarizes the findings of various studies related to lingering oil from the *Exxon Valdez* oil spill. The report estimated that at least 10,800,000 gallons of crude oil spilled from the *Exxon Valdez* into the Prince William Sound in 1989 and that based on data collected from 2001 through 2007, residual oil on the shoreline represented 0.25 percent of the total spill volume—or about 27,000 gallons. The report further stated that there could be any type of subsurface oil on 35 kilometers of shoreline—about 22 miles.

will persist for decades; however, its representatives said that the evidence indicates that there are no current biological effects of the oil.

Source: GAO. | GAO-19-31.

Figure 3. A Researcher Collects Lingering Oil from the Exxon Valdez Oil Spill on a Beach in Alaska in May 2018.

The Exxon Valdez Trustee Council's priorities for future spending are outlined in the 2014 restoration plan update, and in addition to long-term herring research and lingering oil, the priorities include long-term monitoring of marine conditions and injured resources, shorter-term harbor restoration projects, and habitat protection.

The Deepwater Horizon Trustee Council Has Used 13 Percent of the Restoration Trust Fund, and Most Restoration Activities Are in the Initial Planning Phase

Since the federal and state governments reached a final settlement with BP in 2016 and the Deepwater Horizon Trustee Council finalized a programmatic restoration plan, four trustee implementation groups have issued initial independent restoration plans. Specifically, the Alabama, Louisiana, Mississippi, and Texas trustee implementation groups have issued initial restoration plans. According to the Deepwater Horizon Trustee Council, the trustee implementation groups covering Florida, Open Ocean, and Region-Wide restoration are in the midst of a multiyear planning effort and anticipate issuing initial restoration plans in 2019 or later. The trustee implementation groups are responsible for developing and approving restoration plans and resolutions, which, when approved, authorize money to be spent on restoration projects. This process includes soliciting project ideas, submitting proposed plans for public comment, and ensuring compliance with applicable laws and regulations, such as the National Environmental Policy Act. According to the trustee council, there is no specific timetable for approving future restoration plans, as plans are approved on an ongoing basis—typically for several projects at a time. The four completed restoration plans, together with early restoration spending and other activities, including planning and administrative efforts, account for all authorizations made by the Deepwater Horizon Trustee Council as of December 31, 2017, according to NOAA—the agency that manages the system the trustee councils uses for financial reporting.[40] As shown in Figure 4, these authorizations include approximately $1.1 billion, or 13 percent, of the $8.1 billion restoration trust fund on five goals.[41]

The Deepwater Horizon Trustee Council has authorized roughly $460 million for habitat protection—about 10 percent of the almost $4.7 billion ordered for this use by the settlement.[42] According to the 2016 programmatic restoration plan, habitat protection includes both

---

[40] According to the federal trustees, the trustee council authorizes funds for multiyear restoration activities up front. Therefore, expenditures accrue over subsequent years for these restoration activities. As of the end of 2017, the trustees had spent $368 million, or 5 percent of the restoration trust fund.

[41] The Deepwater Horizon restoration trust fund had received approximately $1.5 billion in payments from BP as of December 31, 2017. Therefore, the Deepwater Horizon Trustee Council had authorized about 73 percent of the funds received from BP as of the end of 2017. As previously noted, the settlement also orders up to $700 million for unknown conditions and adaptive management, which is not included in this $8.1 billion or the following percentage calculations because the funds cannot be demanded before January 1, 2026.

[42] The total amount of the natural resource damage payments ordered for each restoration goal is defined in the consent decree.

conservation acquisition and habitat enhancement, such as creating, restoring, or enhancing coastal wetlands. For example, during the first phase of early restoration in 2012, the trustee council authorized $14.4 million to the Louisiana Coastal Protection and Restoration Authority to create 104 acres of new brackish marsh at Lake Hermitage in Barataria Bay, Louisiana. The project involved dredging sediment and planting native marsh vegetation to restore marsh habitat damaged by the spill. The project is currently in the monitoring phase. As of the end of 2017, the Deepwater Horizon Trustee Council had approved 34 habitat protection projects, many of which were still in progress as of December 2017. The initial results of these projects include the restoration of over 4,000 acres of habitat and the creation of over 40 artificial reefs, according to a written statement by the federal trustees.

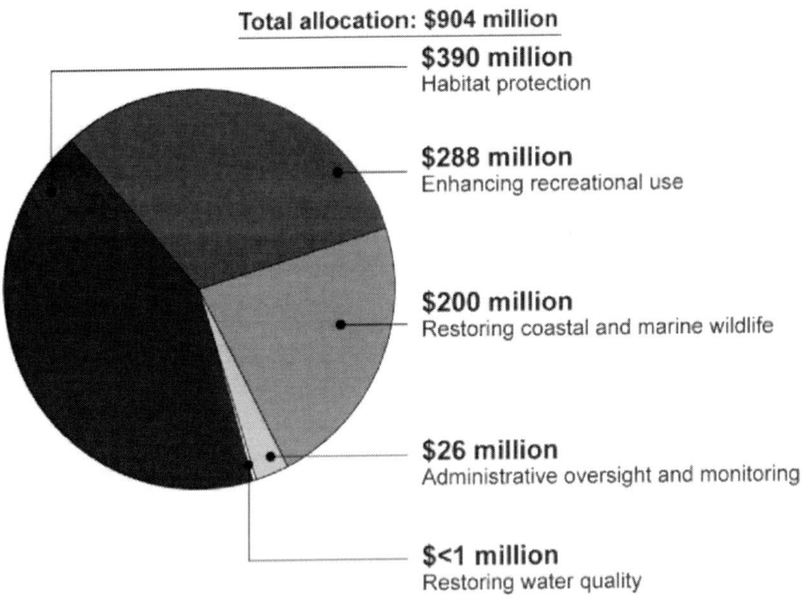

Source: GAO analysis of Deepwater Horizon Natural Resource Damage Assessment Trustee Council data.

Figure 4. Deepwater Horizon Natural Resource Damage Assessment Trustee Council Use of Restoration Trust Fund by Restoration Goal, from April 2012 to December 31, 2017.

The trustee council has authorized roughly $349 million to enhance recreational use—about 83 percent of the almost $420 million ordered for this use by the settlement. According to the 2016 programmatic restoration plan, enhancing recreational use includes acquiring land along the coast, building improved or new infrastructure, and improving navigation for on-water recreation. For example, during the first phase of early restoration in 2012, the Deepwater Horizon Trustee Council authorized approximately $5.3 million to the Florida Department of Environmental Protection to repair and construct boat ramps in Pensacola Bay and Perdido Bay, Florida. Construction was completed in 2016, and the project is currently in the monitoring and operations and maintenance phase. As of the end of 2017, the Deepwater Horizon Trustee Council had approved 43 projects to enhance recreational use, many of which were still in progress as of December 2017. These projects have provided new or enhanced facilities, such as pavilions, picnic areas, and boat ramps, according to a written statement by the federal trustees.

The Deepwater Horizon Trustee Council has authorized roughly $218 million to restore coastal and marine wildlife—about 12 percent of the almost $1.8 billion ordered for this use by the settlement, primarily for birds ($108 million), sea turtles ($50 million), oysters ($38 million), and fish ($20 million). According to the 2016 programmatic restoration plan, restoring coastal and marine wildlife includes activities that restore the resources, such as fish, sea turtles, and deep coral communities, which contribute to a productive, biologically diverse, and resilient ecosystem. For example, during the first phase of early restoration in 2012, the trustee council authorized $11 million to the Mississippi Department of Environmental Quality to deploy a mixture of oyster shells, limestone, and concrete on 1,430 acres in waters off Hancock and Harrison Counties in Mississippi. This material, when placed in oyster spawning areas, provides a surface for free swimming oyster larvae to attach and grow into oysters. The project is currently in the monitoring and operations and maintenance phase. As of the end of 2017, the Deepwater Horizon Trustee Council had approved 32 projects to restore coastal and marine wildlife.

Although the trustee council authorized millions of dollars to restore coastal and marine wildlife, it authorized 1 percent or less of funds ordered by the settlement for sturgeon, marine mammals, submerged aquatic vegetation, and other seafloor species—such as corals.[43] According to the 2016 consent decree, the Open Ocean implementation group is responsible for authorizing the majority of the restoration funds for these types of wildlife, but that trustee implementation group has not yet completed its initial restoration plan. According to NOAA, the complexity of restoring several of these resources necessitated additional preplanning and restoration technique development prior to considering specific restoration projects for several of these types of wildlife. The trustee implementation group is developing two restoration plans that will include projects for birds and sturgeon, as well as for sea turtles, fish, marine mammals, and corals, according to a Deepwater Horizon Trustee Council press release. The trustee council released the first draft plan for public comment in October 2018, and plans to release the second plan in early 2019. In August 2017, the Deepwater Horizon Trustee Council announced that the Louisiana implementation group was soliciting project ideas to fund the restoration of submerged aquatic vegetation, among other types, to include in a future restoration plan but has not yet submitted such a plan for public review.

Roughly $27 million has been authorized for administrative oversight and monitoring activities, or about 3 percent of the almost $810 million that the settlement ordered for this use. The majority of the funding ($25 million) was for administrative oversight activities, and the balance was for monitoring. According to the 2016 programmatic restoration plan, administrative oversight includes the costs for trustees to guide project selection, implementation, and adaptive management. For the state trustees, all administrative costs are covered by their respective trustee implementation groups, and for federal trustees, all administrative costs are

---

[43] Corals are an example of a species covered within the type called mesophotic and deep benthic communities.

covered by the Open Ocean implementation group.[44] For example, during the post-settlement phase, the trustee council authorized approximately $6.6 million to Interior for (1) participation on the trustee council; (2) restoration planning, plan development, and coordination with other trustees; (3) environmental compliance reviews; (4) technical assistance; and (5) financial management, among other uses. As of the end of 2017, the Deepwater Horizon Trustee Council had approved nine administrative oversight and monitoring projects, which remained ongoing as of December 31, 2017. The results of the trustee council's activities in this area so far include the completion of a monitoring and adaptive management manual and its standard operating procedures.[45]

The Deepwater Horizon Trustee Council has authorized $4 million to restore water quality—about 1 percent of the $410 million that the settlement ordered for this use. According to the 2016 programmatic restoration plan, restoring water quality includes both reducing nonpoint nutrient pollution to coastal watersheds and improving water quality in Florida through efforts such as stormwater control and erosion control. As of the end of 2017, the Deepwater Horizon Trustee Council approved two nonpoint nutrient reduction projects to address excessive nutrient loads in Gulf waters but no water quality projects in Florida.[46]

---

[44] The exception is costs for the Lead Administrative Trustee functions for the designated federal agency, which are authorized by the Region-Wide implementation group. In addition, according to the consent decree, the Region-Wide restoration area contains funding for Gulf-wide needs, such as monitoring, oversight, planning, and research.

[45] Deepwater Horizon Natural Resource Damage Assessment Trustees, *Monitoring and Adaptive Management Procedures and Guidelines Manual Version 1.0* (December 2017), and *Trustee Council Standard Operating Procedures for Implementation of the Natural Resource Restoration for the Deepwater Horizon Oil Spill* (Nov. 15, 2016).

[46] Nutrient pollution, caused by excess nitrogen and phosphorus entering the environment, poses significant risks to the nation's water quality and presents a growing threat to public health and local economies. Excess nutrients can enter water bodies from point sources, such as industrial facilities and wastewater treatment plans, and from nonpoint sources, such as runoff from agricultural and urban areas. Nutrient pollution contributes to a trend of increasing numbers of harmful algal blooms in surface waters, which can release toxins that can contaminate drinking water sources and adversely affect recreation, tourism, and fisheries. See GAO, *Water Pollution: Some States Have Trading Programs to Help Address Nutrient Pollution, but Use Has Been Limited*, GAO-18-84 (Washington, D.C.: Oct. 16, 2017).

For example, in 2017, the Deepwater Horizon Trustee Council authorized approximately $224,000 to conduct restoration planning to develop, draft, and finalize a restoration plan addressing nonpoint nutrient reduction, among other goals. The trustee council has authorized few funds to date for this restoration goal because, in part, the Florida implementation group has not yet completed its first postsettlement restoration plan. In September 2017, the trustee council announced that the Florida implementation group was reviewing water quality project ideas for its initial restoration plan, and it released a draft of the plan for public comment in September 2018. According to the Deepwater Horizon Trustee Council, the final plan will be released in January 2019.

Interagency Committee Members Funded Oil Spill Research Projects from Fiscal Years 2011 through 2017, but the Committee Did Not Coordinate with All Relevant Entities

Nine of the interagency committee member agencies funded over 100 oil spill research projects per year from fiscal years 2011 through 2017, for a total cost of about $200 million; however, we found that the interagency committee did not coordinate its research with some key entities.[47] More specifically, approximately half of the interagency committee members said internal coordination on such research improved during this time, but the committee may not have included all relevant agencies, and we found that the committee did not coordinate with relevant trustee councils.

[47] The nine interagency committee member agencies reported that many of the oil spill research projects they reported funding are multiyear projects; therefore, they reported ongoing projects for each fiscal year rather than those initiated or completed in a single year. This prevented us from providing an overall number of projects by agency from fiscal years 2011 through 2017.

Nine Member Agencies Funded over 100 Oil Spill Research Projects per Year for a Cost of About $200 Million from Fiscal Years 2011 through 2017

During fiscal years 2011 through 2017, 9 of the 15 interagency committee member agencies funded oil spill research projects, spending about $200 million on this research, based on our review of agency data from the member agencies.[48] These nine agencies were the Bureau of Ocean Energy Management, BSEE, the Coast Guard, the Department of Energy, EPA, NASA, NOAA, the Pipeline and Hazardous Materials Safety Administration, and the U.S. Arctic Research Commission. One of these agencies—BSEE—spent about $84 million, or about 40 percent of the total amount spent by all nine agencies (see Table 1).

In March 2011 we reported that during fiscal years 2000 through 2010, seven interagency committee member agencies spent about $163 million on oil pollution research, according to officials from those agencies. Since we last reported on the interagency committee, three additional agencies told us that they also fund oil spill research—the Department of Energy, BSEE,[49] and the U.S. Arctic Research Commission—while the U.S. Navy told us that it no longer funds oil spill research projects.[50]

---

[48] The nine member agencies that reported funding oil spill research during the time period covering fiscal years 2011 through 2017 are listed in table 1. However, NOAA was unable to provide reliable data on the agency's expenditures so we did not include its estimate in the total provided. Six member agencies reported that they did not fund or conduct oil spill research from fiscal years 2011 through 2017. The Department of Commerce's National Institutes of Standards and Technology, the Department of Defense's U.S. Army Corps of Engineers, the Department of Homeland Security's U.S. Fire Administration (within the Federal Emergency Management Agency), Interior's U.S. Fish and Wildlife Service, the U.S. Navy, and the Department of Transportation's Maritime Administration all reported that they did not fund oil spill research using agency funds during this time period.

[49] In October 2011, Interior created two new, independent agencies—the Bureau of Ocean Energy Management and the Bureau of Safety and Environmental Enforcement—to carry out the offshore energy management and enforcement functions of the previously established Bureau of Ocean Energy Management, Regulation, and Enforcement. In our previous report, this bureau had reported funding $77.8 million for oil pollution research projects during fiscal years 2000 through 2010. GAO-11-319.

[50] The U.S. Arctic Research Commission requested to be added as a member of the interagency committee. Although the President is authorized to administratively designate additional federal agencies as members of the interagency committee, OPA was amended in 2016 to

Table 1. Interagency Coordinating Committee on Oil Pollution Research Member Agencies' Expenditures on Oil Spill Research, Fiscal Years 2011 through 2017

	Fiscal year (dollars in millions)							**Total (dollars in million)**
Member agency	2011	2012	2013	2014	2015	2016	2017	
Bureau of Ocean Energy Management	4.3	4.6	3.9	5.5	4.7	4.9	2.5	30.4
Bureau of Safety and Environmental Enforcement (BSEE)[a]	1.1	1.7	8.0	8.1	23.1	23.5	18.0	83.5
Coast Guard	0.8	2.1	1.8	0.2	1.2	1.8	1.5	9.4
Department of Energy	11.8	17.2	3.5	3.8	2.3	2.0	0.8	41.4
Environmental Protection Agency (EPA)	1.1	0.6	0.4	0.7	0.6	0.7	0.2	4.3
National Aeronautics and Space Administration (NASA)	0	0.6	0.4	0.4	0.2	0.2	0.1	1.9
National Oceanic and Atmospheric Administration (NOAA)[b]	NA	NA	NA	NA	NA	NA	NA	NA
Pipeline and Hazardous Materials Safety Administration	5.5	3.2	2.5	2.3	3.1	5.6	5.3	27.5
U.S. Arctic Research Commission[c]	0	NI	$0	$0	$0	$0	NI	NI
Total	24.6	30.0	20.5	21.0	35.2	38.7	28.4	198.4[d]

Legend: NA = data not available; NI = data not included.

Source: GAO analysis of interagency committee member agency data. | GAO-19-31

Notes: Agencies reported spending funds from agency appropriations, as well as appropriations from the Oil Spill Liability Trust Fund. Each agency included research expenditures based on the Interagency Coordinating Committee on Oil Pollution Research's definition and categorization of oil spill research.

[a]BSEE officials reported expending about $1.7 million to $4 million per fiscal year for the operation and maintenance of the Ohmsett National Oil Spill Response Research and Renewable Energy Test Facility located in Leonardo, New Jersey. This facility is used for both full-scale equipment testing and responder training to support oil spill research. These expenditures are not included in the table because agency officials said these funds are used to maintain the facility and not to directly support oil spill research projects.

[b]NOAA officials said they believe that the agency spent about $1 million to $2 million per fiscal year in this period but were unable to provide reliable data on the agency's expenditures because there is no centralized funding or a tracking mechanism for oil spill research.

[c]The U.S. Arctic Research Commission reported expenditures of $65,000 in fiscal year 2012 and $20,000 in fiscal year 2017, but because of rounding, these expenditures would appear as $0.0 in the table above, so we did not include them. For the other fiscal years in the table, the commission reported that it did not fund any oil spill research.

[d]The total does not include the yearly expenditures from NOAA because the agency was not able to provide reliable data on its expenditures.

---

include the commission as a member. Pub. L. No. 114-120, Title III, § 319, 130 Stat. 66 (2016).

According to agency officials, the nine interagency committee member agencies funded from 100 to 200 research projects annually from fiscal years 2011 through 2017. These nine agencies reported funding research projects in one or more of the interagency committee's four oil spill research categories: prevention, preparedness, response, and injury assessment and restoration (see Table 2).

Table 2. Interagency Coordinating Committee on Oil Pollution Research Member Agencies' Oil Spill Research by Category, Fiscal Years 2011 through 2017

	Interagency committee research category				
Member agency	Prevention	Preparedness	Response	Injury assessment and restoration	Cross cutting
Bureau of Ocean Energy Management	-	X	X	-	X
Bureau of Safety and Environmental Enforcement (BSEE)[a]	X	X	X	X	-
Coast Guard[b]	-	X	X	-	-
Department of Energy	X	-	-	-	-
Environmental Protection Agency (EPA)	-	-	X	X	X
National Aeronautics and Space Administration (NASA)	-	-	-	X	X

Source: GAO analysis of interagency committee member agency data. | GAO-19-31.

Notes: The interagency committee organizes research in four categories for its member agencies: (1) prevention, (2) preparedness, (3) response, and (4) injury assessment and restoration. GAO added a fifth category—cross-cutting—to capture those projects that may address more than one category. Each agency included research projects based on the interagency committee's general definition and categorization of oil spill research.

[a]BSEE officials noted that they funded one research project that fell under the injury assessment and restoration category during this time frame and said that restoration is not the focus of the agency's oil spill research efforts.

[b]Coast Guard officials said they also conduct research in the prevention research category that could be relevant to oil spill research but is not specifically classified as oil spill research.

The Interagency Committee Improved Internal Research Coordination Efforts but May Not Have Included All Relevant Agencies and Did Not Include the NRDA Trustee Councils

We reported in March 2011 that federal agencies conducted oil pollution research but that the interagency committee had taken limited actions to foster the communication and coordination of this research among member agencies and nonfederal stakeholders.[51] More specifically, we noted that member agencies were not consistently represented on the interagency committee and interested nonfederal stakeholders reported limited contact with the interagency committee. We recommended, among other things, that the Commandant of the Coast Guard direct the chair of the interagency committee, in coordination with member agencies, establish a more systematic process to identify and consult with key nonfederal stakeholders.

Officials from 8 of the 15 member agencies said they believe that the interagency committee's coordination efforts have improved since the *Deepwater Horizon* oil spill in 2010.[52] In response to our recommendation on coordination with nonfederal stakeholders, we found that members consistently attend major oil spill conferences and workshops. In addition, we observed that the interagency committee invites outside speakers and researchers to its meetings to update the membership on ongoing research activities in academia, industry, and the government. The committee charter calls for meetings at least semiannually, but since fiscal year 2011 the interagency committee has held quarterly meetings with member agencies as well as meetings with outside groups of knowledgeable stakeholders.[53]

---

[51] GAO-11-319.

[52] Of the remaining seven member agencies, six said they either had no context to judge the changes (e.g., they were not members of the interagency committee at the time of the *Deepwater Horizon* spill) or they were neutral regarding the changes since 2010. One agency did not answer the question.

[53] Interagency committee member officials noted that they hold meetings with outside groups such as the Prince William Sound Regional Citizens' Advisory Council, the Gulf of Mexico Restoration Initiative, the National Academies of Sciences Gulf Research

At the meetings, member agencies have the opportunity to present information on oil spill research they are conducting, share information about upcoming research conferences, and listen to presentations by outside groups. According to member agency officials, some of the benefits of the interagency committee's improved coordination efforts include

- a reduction in research redundancies,
- increased communication,
- increased understanding of the broader oil spill research community,
- the facilitation of relationships,
- the identification of research gaps, and
- the ability to leverage resources.

U.S. Navy officials said that the interagency committee facilitated communication between member agencies that use the Navy's equipment for research purposes. As a result of discussions that took place at an interagency meeting, the Navy offered the use of a hydraulic power unit to the Coast Guard for hydraulic testing in Arctic conditions in Alaska. Officials from a few of the member agencies, including the Coast Guard, BSEE, EPA, and NOAA, told us that they collaborate on oil spill-related research efforts with other member agencies of the interagency committee.

In addition, the release of the 2015-2021 research and technology plan provides a new baseline for research, including 150 priority oil pollution research needs within 25 research areas. According to the research and technology plan, future updates will reflect advancements in oil pollution technology and changing research needs by capitalizing on the unique roles and responsibilities of each member agency. According to officials from one member agency, the revised research and technology plan has helped member agencies coordinate with other member agencies to leverage

---

Program, industry groups, and academia to encourage coordination of oil spill research efforts.

funding and expertise. Member agencies also cooperate with nonfederal research entities on research needs and activities.

The interagency committee has demonstrated key practices that strengthen coordination, such as agreeing on common terminology and priorities for oil spill research in its revised research and technology plan. However, the committee could enhance coordination by ensuring that relevant participants have been included—another key practice.[54] Under OPA, certain federal agencies are members of the interagency committee, but member agencies may choose which office or official represents them at meetings and coordinates with other members on committee-related work. Officials from 6 of the 15 member agencies told us that their particular research efforts are not the focus of ICCOPR meetings, and therefore ICCOPR's ability to coordinate their research efforts are less valuable. For example, NASA officials said the office representing their agency at meetings is not involved in oil spill research, but other offices within their agency fund or conduct relevant research.

In addition, 7 of the 15 officials we interviewed from member agencies suggested that other federal agencies could be relevant to the committee's research efforts. For example, officials we interviewed from several member agencies suggested including the U.S. Geological Survey (USGS) as a full member because of its relevant research and mapping expertise.[55] According to committee documents, the interagency committee considered adding USGS in 2015 but has not made a decision on USGS's membership. The Commandant of the Coast Guard, under his or her capacity as chair of the interagency committee, has been delegated authority to appoint additional agencies to the committee as appropriate.[56]

A leading practice for collaboration calls for interagency groups to ensure that all relevant participants have been included in collaborative

---

[54] GAO-12-1022.

[55] According to the interagency committee's research and technology plan, USGS participates in the committee as a nonvoting observer. However, the USGS officials do not regularly attend the quarterly meetings.

[56] In Department of Homeland Security Delegation No. 0170.1, II.80, the Secretary delegated to the Commandant of the Coast Guard the authority to carry out the functions in Executive Order 12777, including § 8(h).

efforts.[57] According to this leading practice, participants should have the appropriate knowledge, skills, and abilities to contribute to the outcomes of the collaborative effort.

However, interagency committee member agency officials said the committee has not systematically reviewed its membership to determine which offices within current member agencies are the most relevant to its mission and whether adding other federal agencies as members would be beneficial. By systematically reviewing its membership to determine whether any additional agencies should be involved in coordinating oil spill research and that the most appropriate offices within member agencies are represented, the interagency committee could improve its ability to coordinate research among federal agencies.

In addition, agency officials knowledgeable about the work of the NRDA trustee councils are not the same officials representing their agency as members on the interagency committee. The research and technology plan notes that the interagency committee's injury assessment and restoration research is intended to support the NRDA process. However, the NRDA trustees who manage the restoration funds for the *Exxon Valdez* and *Deepwater Horizon* oil spills told us that they have not coordinated or communicated on oil spill research or restoration efforts with the interagency committee; therefore, they would not have been involved with developing the research and technology plan. In addition, some trustee council members told us that they were not even aware that the interagency committee existed.

Under OPA, one of the interagency committee's responsibilities is to coordinate with federal agencies and external entities on an oil pollution research program that includes methods to restore and rehabilitate natural resources damaged by oil spills. As previously discussed, the NRDA trustee councils are charged with assessing natural resource damages for the natural resources under their trusteeship and developing and implementing plans for restoration efforts. The research that the interagency committee members fund includes research on restoration that could be pertinent to the

---

[57] GAO-12-1022.

work of the NRDA trustee councils. For example, following the oil spill in 2010, the Deepwater Horizon Trustee Council evaluated baseline conditions for several different representative species, such as sea turtles and Gulf sturgeon, to quantify the extent of injury as part of the restoration planning process that OPA regulations required. Some interagency committee member agencies, such as NOAA and BOEM, fund research on baseline data that could inform the NRDA trustee councils' injury assessment work. In turn, the NRDA trustee councils' work could also inform the interagency committee's coordination of future oil spill research by, for example, identifying gaps in research as identified and prioritized in updates to the research and technology plan. By coordinating with the NRDA trustee councils, the interagency committee could ensure that its research informs and supports the councils' damage assessment and restoration efforts and better leverages members' resources.

Literature Suggests the Effectiveness of Offshore Oil Spill Response Techniques Varies Based on Regional Environmental Differences and Other Factors

According to the literature we reviewed, environmental differences between the Gulf of Mexico and Arctic regions, as well as factors such as the type of oil, influence the potential effectiveness of various oil spill response techniques. In each region, environmental conditions, such as water and air temperature, water movement, and salinity, influence how effective oil spill response techniques can be. Further, according to the literature we reviewed, these conditions determine which response techniques are appropriate.

Environmental conditions, such as ocean water and air temperature, can influence the effectiveness of natural oil removal through evaporation or biodegradation.[58] These processes may occur more quickly in warmer climates, such as in the Gulf of Mexico. In the event of an oil spill, communities of microbes can bloom to respond to the new supply of oil. According to a 2011 report from the American Academy of Microbiology, these microbes can biodegrade up to 90 percent of some light crude oil, but the largest and most complex molecules—such as the ones that make up road asphalt—are not significantly biodegradable.[59] A 2016 study found that higher temperatures lead to increased biodegradation, and increased salinity had a small positive impact on crude oil removal.[60] However, the American Academy of Microbiology report also states that while microbes can biodegrade oil over time, the process may not be fast enough to prevent ecological damage. Therefore, immediate containment or physical removal of the oil is an important first response.

The effectiveness of oil removal is also influenced by conditions of the water, determined by wind, waves, and currents. According to literature we reviewed, winds and currents can make it more difficult to remove the oil, increasing the likelihood of the oil spill affecting larger areas and additional plant and animal populations. Further, high seas and rough waters can make some response techniques less effective. According to a 2017 study that estimates the effect of environmental conditions on deploying oil spill response techniques in the Arctic Ocean, most response techniques are not suitable during Arctic winters, between November and June.[61]

Literature we reviewed also shows that other factors influence the effectiveness of response techniques, including oil type, oil thickness, and

---

[58] Biodegradation is the disintegration of materials, such as oil, by bacteria or fungi called microbes. It is estimated that about half of the oil in the world's oceans comes from natural seeps in the oil reservoirs and communities of microbes develop to metabolize this oil for their own energy.

[59] American Academy of Microbiology, *Microbes and Oil Spills FAQ* (Washington, D.C.: 2011).

[60] Priyamvada Sharma and Silke Schiewer, "Assessment of Crude Oil Biodegradation in Arctic Seashore Sediments: Effects of Temperature, Salinity, and Crude Oil Concentration" *Environmental Science and Pollution Research,* vol. 23 (2016).

[61] Sierra Fletcher, Tim Robertson, Bretwood Higman, and Elise DeCola, *Estimating Impact of Environmental Conditions on Deployment of Marine Oil Spill Response Tactics in the U.S. Arctic Ocean*, proceedings of the Fortieth AMOP Technical Seminar (Ottawa: Environment and Climate Change Canada, 2017) 246-264.

the location and depth of oil spill events. Light crude oil typically evaporates and biodegrades more quickly than heavy crude oil, which is more viscous. However, if the oil slick is too thin, it becomes difficult to contain and limits response options. Oil spilled in a remote location, such as the place where the *Exxon Valdez* oil spill occurred, may complicate response efforts because equipment and personnel are far away and may not be able to respond within the window of opportunity before the oil spreads.

According to Coast Guard officials, during an oil spill response, various response techniques are used to minimize the negative effects on the water surface, water column, and shorelines, each with different applications, advantages, disadvantages, and risks. The response techniques we reviewed are:

- Mechanical recovery in the marine environment uses a variety of containment booms, barriers, and skimmers, as well as natural and synthetic absorbent materials to capture and store the spilled oil until it can be disposed of properly.
- In-situ burning, meaning in-place burning, is the process of igniting and burning oil slicks in a controlled environment.
- Dispersants are chemicals that can mitigate the immediate damage caused by oil at the surface and help accelerate the natural removal of the spilled oil. Dispersants work similarly to dish soap, breaking up the oil into small droplets that can more easily spread through the water.[62]

The advantage of mechanical recovery is that it physically removes the oil from the water, minimizing the negative effects of the oil. Mechanical recovery can be used to safely remove oil where other methods might

[62] Separating the oil molecules in oil slicks helps the oil move away from the surface down toward the ocean floor, which can accelerate the ocean's process of weathering or biodegrading the oil.

cause health risks or environmental damage, according to a 2013 report published by the National Academies Press.[63]

Mechanical Recovery Safely Removes Spilled Oil but Has Limitations in Certain Conditions

However, mechanical recovery has limitations in some conditions. If the oil slick is thin, it is difficult to achieve a significant rate of recovery and requires a lot of equipment to concentrate the slick so it is thick enough to be collected. According to literature we reviewed, mechanical recovery is less effective during inclement weather or high seas because the oil spreads and can emulsify in these conditions and is difficult to contain. Low temperatures and the presence of ice also make it challenging to achieve high recovery rates, and mechanical recovery becomes increasingly ineffective as wave heights increase, according to literature we reviewed.

Furthermore, the process of recovering the oil is labor- and cost-intensive, and recovery can be delayed if the equipment is not readily available. Mechanical recovery is especially challenging to implement quickly when spills occur in remote areas, such as with *Exxon Valdez*, or where the oil is traveling quickly and broadly, such as with *Deepwater Horizon*. For example, according to a 1999 EPA report, skimmers were not readily available during the first 24 hours following the *Exxon Valdez* oil spill, repairs to damaged skimmers were time-consuming, and continued inclement weather slowed down the recovery efforts.[64] In addition, a disadvantage of mechanical recovery is that temporary storage for large amounts of oil is frequently needed and recovered oil is generally brought

---

[63] Committee on the Effects of the *Deepwater Horizon* Mississippi Canyon-252 Oil Spill on Ecosystem Services in the Gulf of Mexico, Ocean Studies Board, Division on Earth and Life Studies, and National Research Council, *An Ecosystem Services Approach to Assessing the Impacts of the Deepwater Horizon Oil Spill in the Gulf of Mexico* (Washington, D.C.: National Academies Press, Dec. 20, 2013).

[64] Environmental Protection Agency, Office of Emergency and Remedial Response, *Understanding Oil Spills and Oil Spill Response,* EPA 540-K-99-007 (December 1999), 39.

back to the shore for disposal, according to Interior officials. Because of the resources required to physically remove the oil, it is difficult to recover a large percentage of the spilled oil through mechanical recovery in large oil spills.

In-Situ Burning Can Efficiently Eliminate Oil but Has Potential Side Effects

According to two studies and an agency document we reviewed, in-situ burning can be a highly effective technique for eliminating spilled oil from the sea surface.[65] In response to the *Deepwater Horizon* oil spill, roughly 5 to 6 percent of all of the spilled oil was burned, about double the amount of oil removed with skimmers, according to a 2013 National Academies Press report.[66]

The primary advantage of in-situ burning is its efficiency. In ideal conditions, this method can quickly eliminate spilled oil. According to several reports we reviewed, in optimal conditions, in-situ burning can eliminate up to 90 percent of the spilled oil contained for burning with a relatively minimal investment of equipment or manpower.[67] Literature we reviewed suggests that it is especially suited for response in Arctic conditions, particularly in ice-covered water where logistics and

---

[65] C. C. Guenette and P. Sveum. "In-situ Burning of Uncontained Crude Oil and Emulsion," *Proceedings of the 18th Arctic and Marine Oil Spill Program (AMOP) Technical Seminar, Environment Canada, vol. 2* (1995); A.Y. Walavalkar and A.K. Kulkarni, "Comprehensive Review of Oil Spill Combustion Studies," *Proceedings of the 9th Arctic and Marine Oil Spill Program (AMOP) Technical Seminar, Environment Canada* (1996), 1081–1103.

[66] Committee on the Effects of the *Deepwater Horizon* Mississippi Canyon-252 Oil Spill on Ecosystem Services in the Gulf of Mexico, *An Ecosystem Services Approach to Assessing the Impacts of the Deepwater Horizon Oil Spill in the Gulf of Mexico*.

[67] Janne Fritt-Rasmussen, Susse Wegeberg, and Kim Gustavson, "Review on Burn Residues from In Situ Burning of Oil Spills in Relation to Arctic Waters," *Water Air Soil Pollution,* vol. 226 (2015); Committee on Responding to Oil Spills in the U.S. Arctic Marine Environment, Ocean Studies Board, Polar Research Board, Division on Earth and Life Studies, Marine Board, Transportation Research Board, and National Research Council, *Responding to Oil Spills in U.S. Arctic Marine Environment* (Washington, D.C.: National Academies Press, 2014), 6, 88; and Arctic Oil Spill Response Technology Joint Industry Programme, *Synthesis Report,* D. Dickens-DF Dickens Associates, LLC (May 3, 2017) 39.

environmental conditions may preclude other options and where the ice can act as a natural barrier to help keep the oil slick thick enough to burn.

However, in-situ burning also has its disadvantages. Burning has a narrow window of opportunity, and if the approval process takes longer than it takes to prepare for the burn, the opportunity for using in-situ burning may be lost, according to a NOAA document. Similar to mechanical recovery, burning can only be used if the oil slick is a certain thickness and when waves, wind, and currents are not too strong. In-situ burning becomes increasingly difficult in strong winds or with waves over 3 feet tall. A second disadvantage is that the burn residue caused by in-situ burning may have negative effects on ocean life, though studies we reviewed differed on this matter. According to a 2014 National Academies Press report about oil spills in the U.S. Arctic environment, a series of studies in the 1990s found that burn residues have little to no impact on oceanic organisms.[68] However, a 2015 review on burn residues from in-situ burning in Arctic waters concluded that not enough research has been done on the side effects of burn residue from in-situ burning.[69] According to NOAA officials, another disadvantage of in-situ burning is that the soot from inefficient combustion can result in unsightly and unhealthy particulates that may affect any downwind populations before the smoke dissipates.

Use of Dispersants Is Versatile but Its Effectiveness Depends on Several Factors

According to Coast Guard officials, chemical dispersants are typically used in conjunction with mechanical means and are considered when offshore mechanical methods are recognized as inadequate because of the

---

[68] Committee on Responding to Oil Spills in the U.S. Arctic Marine Environment, Ocean Studies Board, Polar Research Board, Division on Earth and Life Studies, Marine Board; Transportation Research Board, and National Research Council, *Responding to Oil Spills in U.S. Arctic Marine Environment*.

[69] Fritt-Rasmussen, Wegeberg, and Gustavson, "Review on Burn Residues from In Situ Burning of Oil Spills in Relation to Arctic Waters."

spill volume, the geographical extent of the slicks, or specific on-scene environmental conditions.

According to the literature we reviewed, an advantage of dispersants is their versatility. Dispersants are not as limited by environmental conditions as other response techniques, and they can be applied on surface or underwater environments. Further, dispersants can be applied through a variety of mechanisms. For example, they can be applied on oil slicks at the water's surface by boats, planes, or helicopters. Dispersants can also be used below the surface, through subsea injection at the site of the spill, as was applied in response to the *Deepwater Horizon* oil spill.

However, the literature suggests that the effectiveness of dispersants depends on many factors, such as the type of oil, type of dispersant used, and sea and weather conditions. According to Coast Guard officials, the decision to use dispersants is made after careful consideration of the location of the spill, type of oil spilled, seasonal resources at risk, and the environmental conditions at the time, as these factors influence the effectiveness and practicality of using dispersants, as well as the advisability of the tactic in the face of other options and risks. These officials also noted that dispersants are rarely used in the United States, but in certain situations, where mechanical means such as booming and skimming may not be effective, dispersants may be considered.

In addition to the uncertainty of their effectiveness, the potential environmental risks associated with dispersants are also uncertain. One 2014 study states that while dispersants were thought to undergo rapid degradation in the water column, there was evidence that the dispersants remained on Gulf of Mexico beaches almost 4 years after the *Deepwater Horizon* oil spill.[70] During the *Deepwater Horizon* oil spill, responders applied over 1.8 million gallons of chemical dispersants to the spilled oil—an unprecedented volume in the United States. It was the first major oil spill to use dispersants on such a large scale,[71] and approximately 42

[70] Helen K. White, Shelby L. Lyons, Sarah J. Harrison, David M. Findley, Yina Liu, and Elizabeth B. Kujawinski, "Long-Term Persistence of Dispersants following the *Deepwater Horizon* Oil Spill," *Environmental Science & Technology Letters* (2014) 295-299.

[71] In response to the *Ixtoc* spill, which occurred from 1979 through 1980 off the coast of Mexico, responders applied 2.7 million gallons of dispersants to the surface of the water.

percent of these dispersants were applied sub-sea in the first operational sub-sea application of this technique.[72] According to Coast Guard officials, the toxicity and long-term effects of large-scale application of dispersants on the ecology of marine life are unknown. According to literature we reviewed, there is evidence that chemically dispersed oil and some dispersant compounds may be toxic to some marine life, especially those in early life stages. Coast Guard officials also said that continued monitoring and further review of scientific research should improve the understanding of the impact of dispersants on mitigating the effects of oil spills as well as their overall environmental impact.

CONCLUSION

Following initial response and cleanup efforts, restoration activities related to a significant offshore oil spill, such as those from *Exxon Valdez* or *Deepwater Horizon*, can endure for decades. Federal agencies of the interagency committee conduct and fund research projects related to preventing, preparing for, responding to, and restoring the environment after oil spills. The interagency committee has improved the coordination of federal oil spill research efforts since the *Deepwater Horizon* oil spill in 2010. However, the interagency committee has not systematically reviewed its membership to determine which offices within current member agencies are the most relevant to its mission and whether adding other federal agencies as members would be beneficial. By systematically reviewing its membership to determine whether any additional agencies should be involved in coordinating oil spill research and that the most appropriate offices within member agencies are represented, the interagency committee could improve its ability to coordinate research among federal agencies.

In addition, the interagency committee does not coordinate with the NRDA trustee councils that manage the large restoration funds and

[72] United States Coast Guard, On Scene Coordinator Report *Deepwater Horizon* Oil Spill, (September 2011).

monitor the restoration of damaged resources after a specific spill, such as the *Exxon Valdez* and *Deepwater Horizon* oil spills. Coordinating with the NRDA trustee councils could help ensure that the interagency committee's oil spill research program is effectively supporting the damage assessment and restoration efforts of the councils, and better knowledge sharing between groups and leveraging its members' oil spill research resources.

RECOMMENDATIONS FOR EXECUTIVE ACTION

We are making the following two recommendations to the Commandant of the U.S. Coast Guard at the Department of Homeland Security:

The Commandant of the U.S. Coast Guard should direct the chair of the Interagency Coordinating Committee on Oil Pollution Research, in coordination with member agencies, to systematically review its membership to determine whether any additional agencies should be involved in coordinating oil spill research and that the most appropriate offices within member agencies are represented. (Recommendation 1)

The Commandant of the U.S. Coast Guard should direct the chair of the Interagency Coordinating Committee on Oil Pollution Research, in coordination with member agencies, to coordinate with the relevant Natural Resource Damage Assessment trustee councils to help ensure that the interagency committee's research informs and supports the councils' damage assessment and restoration efforts. (Recommendation 2)

AGENCY COMMENTS

We provided our draft report to the Department of Agriculture, Department of Commerce, Department of Defense, Department of Energy, Department of Homeland Security, Department of the Interior, Department of Transportation, Environmental Protection Agency, National Aeronautics and Space Administration, and U.S. Arctic Research Commission for review and comment. In comments reprinted in appendix II, the Department of Homeland Security concurred with our recommendations. In addition, the departments of Commerce, Homeland Security, Interior, and EPA provided technical comments, which we incorporated as appropriate.

As agreed with your office, unless you publicly announce the contents of this chapter earlier, we plan no further distribution until 30 days from the report date. At that time, we will send copies of the report to the appropriate congressional committees; the Secretaries of Agriculture, Commerce, Defense, Energy, Homeland Security, the Interior, and Transportation; the Administrators of EPA and NASA; the Executive Director of the U.S. Arctic Research Commission; and other interested parties.

Sincerely yours,

Frank Rusco
Director, Natural Resources and Environment

Appendix I: Objectives, Scope, and Methodology

This chapter examines (1) how the Natural Resource Damage Assessment (NRDA) trustee councils have used the restoration trust funds for the *Exxon Valdez* and *Deepwater Horizon* oil spills and the status of the restoration efforts; (2) the status of the Interagency Coordinating Committee on Oil Pollution Research's (interagency committee) oil spill research efforts and how coordination of such efforts has changed since we last reported on it in March 2011; and (3) what literature suggests about the effectiveness of various oil spill response techniques in the Arctic and the Gulf of Mexico.

To examine how the NRDA trustee councils used the restoration funds from the *Exxon Valdez* oil spill (from October 1992 to January 2018) and the *Deepwater Horizon* oil spill (from April 2012 to December 2017) for restoration and the status of the restoration efforts, we obtained data from each trustee council on the amount of funds (1) ordered by the settlement for each restoration type; (2) authorized by the trustees for, but not yet spent on, restoration activities (authorizations); (3) spent on restoration activities (expenditures); and (4) not yet authorized for restoration activities (remaining balance) through calendar year 2017 for *Deepwater Horizon* and through January 31, 2018, for *Exxon Valdez*.[73] To assess the reliability of the financial data, we reviewed related budget documentation; interviewed knowledgeable council staff about how fund balances are recorded and reported; reviewed the totals for obvious errors and inconsistencies; and reviewed internal control documents, such as a database manual and standard operating procedures. We determined that the data were sufficiently reliable for the purposes of our report.[74]

We examined the approved restoration plans (1994 restoration plan and 2014 restoration plan update for the *Exxon Valdez* oil spill, and the 2016 programmatic damage assessment and restoration plan for the *Deepwater Horizon* oil spill) and, when available, annual reports on restoration activities (1994 through 2018 annual reports for the Exxon Valdez Oil Spill Trustee Council (Exxon Valdez Trustee Council) and 2016 and 2017 annual financial reports for the Deepwater Horizon Natural Resource Damage Assessment Trustee Council (Deepwater Horizon Trustee Council)). We also reviewed project reports and scientific studies that the trustee councils funded to gain a better understanding of the status of restoration of injured natural resources, restoration priorities, activities, and progress made by the trustee councils.[75] We reviewed laws and regulations that provide the legal authority for federal agencies to intervene and respond after an oil spill, such as the Oil Pollution Act of 1990 (OPA),

---

[73] Our review focuses on the role of the NRDA trustee councils; therefore, we excluded the criminal fine and restitution payment for *Exxon Valdez* and the RESTORE Act funding and the criminal payments paid to the National Fish and Wildlife Foundation and other entities for *Deepwater Horizon* from the scope of this review.

[74] The Exxon Valdez Trustee Council has submitted all of its annual financial statements to a third-party audit. Each Deepwater Horizon Trustee Council trustee is responsible for procuring financial audits no less than once every 3 years, with the first audit occurring in 2017, according to the trustee council's standard operating procedures.

[75] Under OPA, damages for injuries to natural resources and services resulting from an oil spill incident must be based on a restoration plan developed after adequate public notice, opportunity for a hearing, and consideration of all public comment.

the Clean Water Act, and NRDA regulations. We met with officials from the Exxon Valdez Trustee Council to discuss the distribution of settlement money for restoration purposes after the *Exxon Valdez* oil spill, and with officials from the Deepwater Horizon Trustee Council, Gulf Coast Ecosystem Restoration Council (RESTORE Council), and the National Fish and Wildlife Foundation to discuss the distribution of settlement money for restoration purposes after the *Deepwater Horizon* oil spill. Additionally, in May 2018, we traveled to multiple locations in the former spill area in Alaska to observe the extent of restoration efforts and ongoing issues. Along with researchers sent by the Exxon Valdez Trustee Council, we excavated three pits that revealed lingering oil about 6 inches below the surface of the beach on Eleanor Island in Prince William Sound. These researchers told us that oil previously uncovered at this location had been linked to the *Exxon Valdez* oil spill. In addition to fieldwork in Alaska, in November 2017 and February 2018, we attended public meetings in Alabama and Louisiana to learn about restoration plans for the Gulf States.

To examine the status of the interagency committee's federal oil spill research efforts and how coordination of such efforts has changed since we last reported on it in March 2011, we requested funding data and project information on oil spill research from all 15 member agencies of the interagency committee.[76] We received data from the 9 member agencies that reported funding oil spill research projects from fiscal years 2011 through 2017.[77] These 9 agencies provided data on agency expenditures on

---

[76] As of November 2018, the 15 member agencies of the interagency committee are the U.S. Coast Guard (Coast Guard), National Oceanic and Atmospheric Administration (NOAA), Environmental Protection Agency (EPA), Bureau of Safety and Environmental Enforcement (BSEE), Bureau of Ocean Energy Management, U.S. Army Corps of Engineers, Pipeline and Hazardous Materials Safety Administration, U.S. Fish and Wildlife Service, U.S. Navy, National Aeronautics and Space Administration (NASA), Department of Energy, U.S. Fire Administration, Maritime Administration, National Institute of Standards and Technology, and U.S. Arctic Research Commission.

[77] The nine member agencies that reported funding oil spill research from fiscal years 2011 through 2017 are the Coast Guard, NOAA, the Department of Energy, the Bureau of Ocean Energy Management, BSEE, the Pipeline and Hazardous Materials Safety Administration, EPA, NASA, and the U.S. Arctic Research Commission. The other six member agencies reported that they did not fund or conduct oil spill research from fiscal years 2011 through 2017.

oil spill research and the research category of any projects funded. [78] We assessed the reliability of the data by reviewing related documentation, interviewing knowledgeable agency officials, and reviewing agency internal controls for each of the 9 member agencies that provided us data about the steps they take to maintain this information. We determined that in most cases the data were sufficiently reliable for the purposes of our report. However, we chose not to provide the National Oceanic and Atmospheric Administration's (NOAA) agency expenditures for oil spill research because NOAA officials were unable to provide reliable data on the actual amount the agency spent on such research during the time period we requested. In addition, some agency officials we interviewed raised the concern that their agencies do not track oil spill research funding and therefore the information they provided on expenditures for such research may not include all relevant efforts that could inform oil spill prevention, preparedness, response, and restoration.

We also interviewed officials from the 15 member agencies to learn about each agency's oil spill research efforts and participation in and coordination through the interagency committee, and compared their coordination practices to one of our federal leading practices for collaboration for interagency groups to evaluate the interagency committee's efforts to coordinate such research.[79] We chose to focus on the collaboration practice pertaining to participants because it appeared to be the most challenging for the interagency committee based on the findings of our previous March 2011 report, the actions taken by the interagency committee to address our recommendations from that report, and our own findings from our research for this chapter.[80] In addition, we reviewed the 2013 interagency committee charter, the committee's most recent biennial

[78] For this chapter we use "funded" to describe money spent on oil spill research projects. We asked member agencies to include the total number of research projects funded from fiscal years 2011 through 2017.

[79] GAO, *Managing for Results: Key Considerations for Implementing Interagency Collaborative Mechanisms*, GAO-12-1022 (Washington, D.C.: Sept. 27, 2012), and *Results-Oriented Government: Practices That Can Help Enhance and Sustain Collaboration among Federal Agencies*, GAO-06-15 (Washington, D.C.: Oct. 21, 2005).

[80] GAO, *Federal Oil and Gas: Interagency Committee Needs to Better Coordinate Research on Oil Pollution Prevention and Response*, GAO-11-319 (Washington, D.C.: Mar. 25, 2011).

reports to Congress covering fiscal years 2008 through 2017, and the committee's third multiyear research and technology plan for fiscal years 2015 through 2021; attended two committee meetings; and reviewed minutes of eight past meetings. We also reviewed OPA's provisions that established and govern the interagency committee's coordination efforts and membership, as well as various related executive documents.

To examine what literature suggests about the effectiveness of various oil spill response techniques in the Arctic and the Gulf of Mexico, we conducted a literature search for studies and articles that analyzed and summarized the effectiveness of various oil spill response techniques in those regions. We identified existing literature from 1989 (the year of the *Exxon Valdez* oil spill) to March 2018 by searching various databases, such as Scopus and ProQuest. We chose to focus on three primary response techniques—mechanical recovery, in-situ burning, and the use of dispersants—used to clean up after offshore oil spills according to knowledgeable stakeholders and the literature we reviewed. The database search produced over 800 results. Our subject matter expert helped the team narrow this list to 50 results, of which we relied on 16 studies and articles that we determined were most relevant to our research objective of determining the effectiveness of various oil spill response techniques in the Arctic and the Gulf of Mexico. Some literature was not included if it was too specific for the scope of our review. Literature published recently, generally within the past 10 years, was considered more relevant. We supplemented the list of studies from these databases with literature from the Congressional Research Service, the National Academies Press, the Environmental Protection Agency (EPA), NOAA, the American Academy of Microbiology, the Arctic Oil Spill Response Joint Industry Programme, and our previous report on oil dispersants. In total, we relied upon 22 literature results to inform the findings of our objective. For a complete list of the literature, see the bibliography. We shared our summary of the literature search findings with agency officials representing some of the interagency committee member agencies. The following agencies responded with comments and we included their perspectives where relevant: the Department of the Interior, EPA, NOAA, and the U.S. Coast

Guard. We did not independently evaluate the effectiveness of these response techniques.

We conducted this performance audit from July 2017 to January 2019 in accordance with generally accepted government auditing standards. Those standards require that we plan and perform the audit to obtain sufficient, appropriate evidence to provide a reasonable basis for our findings and conclusions based on our audit objectives. We believe that the evidence obtained provides a reasonable basis for our findings and conclusions based on our audit objectives.

APPENDIX II: COMMENTS FROM THE DEPARTMENT OF HOMELAND SECURITY

U.S. Department of Homeland Security
Washington, DC 20528

Homeland Security

November 27, 2018

Frank Rusco
Director, Natural Resources and Environment
U.S. Government Accountability Office
441 G Street, NW
Washington, DC 20548

Re: Management Response to Draft Report: GAO-19-31, "OFFSHORE OIL SPILLS: Restoration and Federal Research Efforts Continue, but Opportunities to Improve Coordination Remain"

Dear Mr. Rusco:

Thank you for the opportunity to comment on this draft report. The U.S. Department of Homeland Security (DHS) appreciates the U.S. Government Accountability Office's (GAO) work in planning and conducting its review and issuing this report.

The Department is pleased to note GAO's acknowledgment that the Interagency Committee on Oil Pollution Research Committee (the Committee), has strengthened communication and coordination among member agencies and nonfederal stakeholders since GAO's 2011 review. As the Committee chair, the U.S. Coast Guard remains committed to further improving collaboration on research efforts related to the prevention of, preparedness for, response to, and restoration after oil spills.

The draft report contained two recommendations with which the Department concurs. Attached find our detailed response to each recommendation. Technical comments were previously provided under separate cover.

Again, thank you for the opportunity to review and comment on this draft report. Please feel free to contact me if you have any questions. We look forward to working with you again in the future.

Sincerely,

JIM H. CRUMPACKER, CIA, CFE
Director
Departmental GAO-OIG Liaison Office

Attachment: Management Response to Recommendations Contained in GAO-19-31

GAO recommended that the Commandant of the U.S. Coast Guard:

Recommendation 1: Direct the chair of the Interagency Committee on Oil Pollution Research, in coordination with member agencies, to review its membership to determine whether any additional agencies may be appropriate for coordinating oil spill research and that the most appropriate offices within member agencies are represented.

Response: Concur. As chair of the Committee, the Coast Guard will lead a systematic review of Committee membership. The review will be introduced at the next Committee meeting on December 12, 2018. Based on the review, the Committee will consider and implement adjustments to its membership, as appropriate. Estimated Completion Date (ECD): December 31, 2019.

Recommendation 2: Direct the chair of the Interagency Committee on Oil Pollution Research, in coordination with member agencies, to coordinate with the Natural Resource Damage Assessment [NRDA] trustee councils to help ensure that the interagency committee's research informs and supports the councils' injury assessment and restoration efforts.

Response: Concur. The Committee will coordinate with the NRDA trustee councils to ensure (1) that Committee and councils' members research informs their respective efforts, as appropriate, and (2) develop an action plan for continued engagement. Of note, while Committee research may support the councils' efforts, the Committee will not limit the scope of their Injury and Assessment and Restoration work to areas that only support councils' efforts. ECD: August 31, 2019.

BIBLIOGRAPHY

We reviewed literature to examine what it suggests about the effectiveness of various oil spill response techniques in the Arctic and the Gulf of Mexico. This bibliography contains citations for the studies and articles that contributed to these findings.

American Academy of Microbiology, *Microbes and Oil Spills FAQ* (Washington, D.C.: 2011).

Arctic Oil Spill Response Technology Joint Industry Programme, *Synthesis Report*, D. Dickens-DF Dickens Associates, LLC (May 3, 2017).

Belore, Randy C., Ken Trudel, Joseph V. Mullin, and Alan Guarino. "Large-scale Cold Water Dispersant Effectiveness Experiments with Alaskan Crude Oils and Corexit 9500 and 9527 Dispersants." *Marine Pollution Bulletin*, vol. 58 (2009): 118-128.

Boufadel, Michel C., Xiaolong Geng, and Jeff Short. "Bioremediation of the *Exxon Valdez* Oil in Prince William Sound Beaches." *Marine Pollution Bulletin*, vol. 113 (2016): 156-164.

Brakstad, Odd G., Trond Nordtug, and Mimmi Throne-Holst. "Biodegradation of Dispersed Macondo Oil in Seawater at Low Temperature and Different Oil Droplet Sizes." *Marine Pollution Bulletin*, vol. 93 (2015): 144-152.

Committee on Responding to Oil Spills in the U.S. Arctic Marine Environment; Ocean Studies Board; Polar Research Board; Division on Earth and Life Studies; Marine Board; Transportation Research Board; National Research Council, *Responding to Oil Spills in U.S. Arctic Marine Environment*. National Academies Press (US) (Washington, D.C.: 2014).

Committee on the Effects of the Deepwater Horizon Mississippi Canyon-252 Oil Spill on Ecosystem Services in the Gulf of Mexico, Ocean Studies Board, Division on Earth and Life Studies, National Research Council, *An Ecosystem Services Approach to Assessing the Impacts of the Deepwater Horizon Oil Spill in the Gulf of Mexico*. National Academies Press (US) (Washington, D.C.: December 20, 2013).

Corn, Lynne M., Claudia Copeland, *The Deepwater Horizon Oil Spill: Coastal Wetland and Wildlife Impacts and Response*. Congressional Research Service (July 7, 2010).

Environmental Protection Agency, Office of Emergency and Remedial Response, *Understanding Oil Spills and Oil Spill Response*, EPA 540-K99-007 (Dec 1999).

Fletcher, Sierra, Tim Robertson, Bretwood Higman, and Elise DeCola. *Estimating Impact of Environmental Conditions on Deployment of Marine Oil Spill Response Tactics in the U.S. Arctic Ocean*, proceedings of the Fortieth AMOP Technical Seminar. Ottawa: Environment and Climate Change Canada, 2017, 246-264.

Fritt-Rasmussen, Janne, Susse Wegeberg, d Kim Gustavson, "Review on Burn Residues from In Situ Burning of Oil Spills in Relation to Arctic Waters." *Water Air Soil Pollution*, vol. 226 (2015).

GAO, *Oil Dispersants: Additional Research Needed, Particularly on Subsurface and Arctic Applications*, GAO-12-585 (Washington, D.C.: May 30, 2012).

Naseri, M., and J. Barabady, *Safety and Reliability: Methodology and Applications—Performance of Skimmers in the Arctic Offshore Oil Spills*. London: Taylor & Francis Group, 2015, 607-614.

National Oceanic and Atmospheric Administration, *Oil Spill - Behavior, Response and Planning: Open-water Response Strategies: In-situ Burning*, (August 1997).

Nedwed, Tim, Tom Coolbaugh, and Amy Tidwell. *Subsea Dispersant Use during the Deepwater Horizon Incident*, proceedings of the Thirty-Fifth AMOP Technical Seminar on Environmental Contamination and Response. Vancouver, BC; Canada, ExxonMobil Upstream Research Company, 2012, 506-518.

Nyankson, Emmanuel, Dylan Rodene, and Ram B. Gupta. "Advancements in Crude Oil Spill Remediation Research After the *Deepwater Horizon* Oil Spill." *Water Air Soil Pollution* (2016).

Rahsepar, Shokouh, Martijn P.J. Smit, Albertinka J. Murk, Huub H.M. Rijnaarts, and Alette A.M. Langenhoff. "Chemical Dispersants: Oil

Biodegradation Friend or Foe?" *Marine Pollution Bulletin*, vol. 108 (2016): 113-119.

Ramseur, Jonathan L., *Oil Spills: Background and Governance*. Congressional Research Service (Sept 15, 2017).

Sharma, Priyamvada, and Silke Schiewer. "Assessment of Crude Oil Biodegradation in Arctic Seashore Sediments: Effects of Temperature, Salinity, and Crude Oil Concentration." *Environmental Science and Pollution Research* (2016): 14881-14888.

Shi, X., P.W. Bellino, A. Simeoni, and A.S. Rangwala. "Experimental Study of Burning Behavior of Large-scale Crude Oil Fires in Ice Cavities." *Fire Safety Journal*, vol. 79 (2016): 91-99.

United States Coast Guard, On Scene Coordinator Report *Deepwater Horizon* Oil Spill, (September 2011).

White, Helen K., Shelby L. Lyons, Sarah J. Harrison, David M. Findley, Yina Liu, and Elizabeth B. Kujawinski. "Long-Term Persistence of Dispersants Following the *Deepwater Horizon* Oil Spill." *Environmental Science & Technology Letters* (2014): 295-299.

INDEX

A

B

C

D

E

F

G

H

I

J

L

M

N

O

P

R

S

T

U

V

W

Related Nova Publications

The Activated Sludge Process: Methods and Recent Developments

Editor: Benjamin Lefèbvre

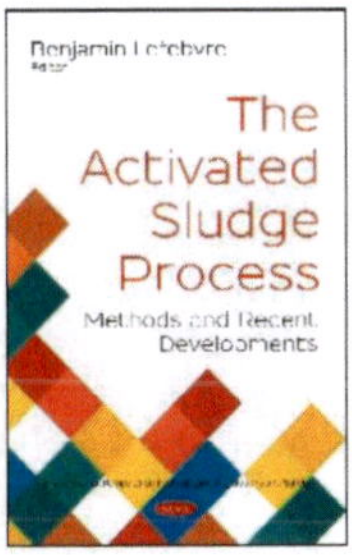

Series: Environmental Remediation Technologies, Regulations and Safety

Book Description: In this collection, the authors report on the pretreatment methods for waste activated sludge based on pulsed electric field and corona discharge techniques. The effects of pulse magnitude, frequency, temperature and pretreatment time are demonstrated on the basis of cell membrane electroporation.

Softcover ISBN: 978-1-53615-202-9
Retail Price: $95

Handbook of Remediation for Complex Environmental Problems

Editors: Sadia Ameen, M. Shaheer Akhtar, and Hyung Shik Shin

Series: Environmental Remediation Technologies, Regulations and Safety

Book Description: In this book, we discuss semiconducting metal oxide nanostructures like TiO2, ZnO, conducting polymers and nanocomposites for their efficient detection of harmful and toxic chemicals, and nanomaterials for photocatalytic degradation, with an emphasis on the applications of semiconducting materials for renewable energy.

Hardcover ISBN: 978-1-53615-118-3
Retail Price: $230

To see a complete list of Nova publications, please visit our website at www.novapublishers.com